the
deepest
well

"*The Deepest Well* . . . offers a powerful — even indispensable — frame to both understand and respond more effectively to our most serious social ills." — DAVID BORNSTEIN, *New York Times*

"Can severe childhood stress cause adult stroke, cancer, Alzheimer's and more? Yes, says Harris, a pediatrician who began researching the biological effects of abuse, divorce and other stressors after treating a boy who stopped growing following a sexual assault. But Harris developed a written screening test and strongly advocates exercise, mindfulness, diet, and talk therapy as remedies. An extraordinary, eye-opening book." — *People,* Book of the Week

"Get the book. Read the book. Share the book . . . *The Deepest Well* is about avoidance, therapy, and healing for the children who have ACE in their lives. The country needs Dr. Burke Harris's book." — ThreeKeyYears.org

"A heartbreaking, world-shaking, revolutionary book. In *The Deepest Well,* Nadine Burke Harris uncovers the once-hidden story of why we are the way we are. And she offers a new set of tools, based in science, that can help each of us heal ourselves, our children, and our world." — PAUL TOUGH, author of *How Children Succeed*

"*The Deepest Well* is more than a riveting medical story — it's a must-read guide for recognizing, understanding, and treating a condition that many will find in our own homes." — *BookPage*

"This ultra-smart and compassionate book delivers revelations about what is really going on — in our bodies, in our families, in our communities — as a result of childhood toxic stress, as well as targeted solutions for individual healing. My adverse childhood experience (ACE) test result is a nine out of ten. When I needed it, one person extended the hand of hope and help to me. It saved me. This book has the power to extend that hand to countless others."
— ASHLEY JUDD, author of *All That Is Bitter and Sweet*

"*The Deepest Well* is a rousing wake-up call, challenging us to reimagine pressing questions of racial and social justice as matters of public health. The research and stories shared in this highly engaging, provocative book prove beyond a reasonable doubt that millions of lives depend on us finally coming to terms with the long-term consequences of childhood adversity and toxic stress."
— MICHELLE ALEXANDER, author of *The New Jim Crow*

"*The Deepest Well* is a heartbreaking, beautiful book about what might be the most important single issue facing our country's disadvantaged populations: the prevalence of childhood trauma. Relying on her work as a compassionate physician and first-class scientist, Burke Harris weaves together groundbreaking research with touching personal stories. The result is a gripping book that should convince everyone that we have a serious problem, and that unless we address it, the losers will be our nation's children."
— J. D. VANCE, author of *Hillbilly Elegy*

"[A] powerful debut . . . in a winning conversational style . . . this important and compassionate book further sounds the alarm over childhood trauma — and what can be done to remedy its effects."
— *Kirkus Reviews,* starred review

"This powerful book brilliantly exposes and explores one of the most critical health issues we face today. Dr. Burke Harris combines a scientist's rigor with a compassionate doctor's heart to paint an unforgettable picture of what is at the center of what ails so many of our communities. Anyone who cares about people who sometimes struggle should read this book."

— BRYAN STEVENSON, author of *Just Mercy*

the
deepest
well

the
deepest well

HEALING the
LONG-TERM EFFECTS
of CHILDHOOD TRAUMA
and ADVERSITY

NADINE BURKE HARRIS, M.D.

MARINER

An Imprint of HarperCollins*Publishers*
Boston   New York

MARINERBOOKS.COM

DESIGNED BY MICHAELA SULLIVAN

Library of Congress Cataloging-in-Publication Data has been applied for.
ISBN 978-1-328-50266-7 (pbk.)
ISBN 978-0-544-82872-8 (e-book)

Printed in the United States of America

23 24 25 26 27 LBC 15 14 13 12 11

CYW Adverse Childhood Experiences Questionnaire reprinted by permission of
Center for Youth Wellness.

To my patients and to the community
of Bayview Hunters Point.
Thank you for teaching me more
than any university possibly could.

Contents

Author's Note

All of the stories in this book are true. Names and identifying details of some individuals have been changed in some circumstances to protect confidentiality. Some vignettes are retold from previous published works.

Introduction

AT FIVE O'CLOCK ON an ordinary Saturday morning, a forty-three-year-old man — we'll call him Evan — wakes up. His wife, Sarah, is breathing softly beside him, curled in her usual position, arm slung over her forehead. Without thinking much about it, Evan tries to roll over and slide out of bed to get to the bathroom, but something's off.

He can't roll over and it feels like his right arm has gone numb.

Ugh, must have slept on it too long, he thinks, bracing himself for those mean, hot tingles you get when the circulation starts again.

He tries to wiggle his fingers to get the blood flowing, but no dice. The aching pressure in his bladder isn't going to wait, though, so he tries again to get up. Nothing happens.

What the . . .

His right leg is still exactly where he left it, despite the fact that he tried to move it the same way he has been moving it all his life — without thinking.

He tries again. Nope.

Looks like this morning, it doesn't want to cooperate. It's weird, this whole body-not-doing-what-you-want-it-to thing, but the urge to pee feels like a much bigger problem right now.

"Hey, baby, can you help me? I gotta pee. Just push me out of bed so I don't do it right here," he says to Sarah, half joking about the last part.

"What's wrong, Evan?" says Sarah, lifting her head and squinting at him. "Evan?"

Her voice rises as she says his name the second time.

He notices she's looking at him with deep concern in her eyes. Her face wears the expression she gets when the boys have fevers or wake up sick in the middle of the night. Which is ridiculous because all he needs is a little push. It's five in the morning, after all. No need for a full-blown conversation.

"Honey, I just gotta go pee," he says.

"What's wrong? Evan? What's wrong?"

In an instant, Sarah is up. She's got the lights on and is peering into Evan's face as though she is reading a shocking headline in the Sunday paper.

"It's all right, baby. I just need to pee. My leg is asleep. Can you help me real quick?" he says.

He figures that maybe if he can put some pressure on his left side, he can shift position and jump-start his circulation. He just needs to get out of the bed.

It is in that moment that he realizes it isn't just the right arm and leg that are numb — it's his face too.

In fact, it's his whole right side.

What is happening to me?

Then Evan feels something warm and wet on his left leg.

He looks down to see his boxers are soaked. Urine is seeping into the bed sheets.

"Oh my God!" Sarah screams. In that instant, seeing her husband wet the bed, Sarah realizes the gravity of the situation and leaps into action. She jumps out of bed and Evan can hear her running to their teenage son's bedroom. There are a few muffled words that he can't make out through the wall and then she's back. She sits on the bed next to him, holding him and caressing his face.

"You're okay," Sarah says. "It's gonna be okay." Her voice is soft and soothing.

"Babe, what's going on?" Evan asks, looking at his wife. As he gazes up at her, it dawns on him that she can't understand anything he's saying. He's moving his lips and words are coming out of his mouth, but she doesn't seem to be getting any of it.

Just then, a ridiculous cartoon commercial with a dancing heart bouncing along to a silly song starts playing in his mind.

F stands for face drooping. *Bounce. Bounce.*

A stands for arm weakness. *Bounce. Bounce.*

S stands for speech difficulty.

T stands for time to call 911. Learn to identify signs of a stroke. Act FAST!

Holy crap!

. . .

Despite the early hour, Evan's son Marcus comes briskly to the doorway and hands his mom the phone. As father and son lock eyes, Evan sees a look of alarm and worry that makes his heart clench in his chest. He tries to tell his son it will be okay, but it's clear from the boy's expression that his attempt at reassurance is only making things worse. Marcus's face contorts with fear, and tears start streaming down his cheeks.

On the phone with the 911 operator, Sarah is clear and forceful.

"I need an ambulance right now, *right now!* My husband is having a stroke. Yes, I'm sure! He can't move his entire right side. Half of his face won't move. No, he can't speak. It's totally garbled. His speech doesn't make any sense. Just hurry up. Please send an ambulance *right away!*"

. . .

The first responders, a team of paramedics, make it there inside of five minutes. They bang on the door and ring the bell. Sarah runs downstairs and lets them in. Their younger son is still in his bedroom asleep, and she's worried that the noise will wake him, but fortunately, he doesn't stir.

Evan stares up at the crown molding and tries to calm down. He feels himself starting to drift off, getting further away from the current moment. *This isn't good.*

The next thing he knows, he is on a stretcher being carried down the stairs. As the paramedics negotiate the landing, they pause to shift positions. In that slice of a second, Evan glances up and catches one of the medics watching him with an expression that makes him go cold. It's a look of recognition and pity. It says, *Poor guy. I've seen this before and it ain't good.*

As they are passing through the doorway, Evan wonders whether he will ever come back to this house. Back to Sarah and his boys. From the way that medic looked at him, Evan thinks the answer might not be yes.

When they get to the emergency room, Sarah is peppered with questions about Evan's medical history. She tells them every detail of Evan's life she thinks might be relevant. He's a computer programmer. He goes mountain biking every weekend. He loves playing basketball with his boys. He's a great dad. He's happy. At his last checkup the doctor said everything looked great. At one point, she overhears one of the doctors relating Evan's case to a colleague over the phone: "Forty-three-year-old male, nonsmoker, no risk factors."

But unbeknownst to Sarah, Evan, and even Evan's doctors, he did have a risk factor. A mighty big one. In fact, Evan was more than twice as likely to have a stroke as a person without this risk factor. What no one in the ER that day knew was that, for decades, an invisible biological process had been at work, one involving Evan's cardiovascular, immune, and endocrine systems. One that might very well have led to the events of this moment. The risk factor and its potential impact never came up in all of the regular checkups Evan had had over the years.

What put Evan at increased risk for waking up with half of his body paralyzed (and for numerous other diseases as well) is not rare. It's something two-thirds of the nation's population is exposed to, something so common it's hiding in plain sight.

So what is it? Lead? Asbestos? Some toxic packing material?

It's childhood adversity.

Most people wouldn't suspect that what happens to them in childhood has anything to do with stroke or heart disease or cancer. But many of us do recognize that when someone experiences childhood trauma, there may be an emotional and psychological impact. For the unlucky (or some say the "weak"), we know what the worst of the fallout looks like: substance abuse, cyclical violence, incarceration, and mental-health problems. But for everyone else, childhood trauma is the bad memory that no one talks about until at least the fifth or sixth date. It's just drama, baggage.

Childhood adversity is a story we think we know.

Children have faced trauma and stress in the form of abuse, neglect, violence, and fear since God was a boy. Parents have been getting trashed, getting arrested, and getting divorced for almost as long. The people who are smart and strong enough are able to rise above the past and triumph through the force of their own will and resilience.

Or are they?

We've all heard the Horatio Alger–like stories about people who have experienced early hardships and have either overcome or, better yet, been made stronger by them. These tales are embedded in Americans' cultural DNA. At best, they paint an incomplete picture of what childhood adversity means for the hundreds of millions of people in the United States (and the billions around the world) who have experienced early life stress. More often, they take on moral overtones, provoking feelings of shame and hopelessness in those who struggle with the lifelong impacts of childhood adversity. But there is a huge part of the story missing.

Twenty years of medical research has shown that childhood adversity literally gets under our skin, changing people in ways that can endure in their bodies for decades. It can tip a child's developmental trajectory and affect physiology. It can trigger chronic inflammation and hormonal changes that can last a lifetime. It can alter the way DNA is read and how cells replicate, and it can dramatically increase the risk for heart disease, stroke, cancer, diabetes — even Alzheimer's.

This new science gives a startling twist to the Horatio Alger tale we think we know so well; as the studies reveal, years later, after having "transcended" adversity in amazing ways, even bootstrap heroes find themselves pulled up short by their biology. Despite rough childhoods, plenty of folks got good grades and went to college and had families. They did what they were supposed to do. They overcame adversity and went on to build successful lives — and then they got sick. They had strokes. Or got lung cancer, or developed heart disease, or sank into depression. Since they hadn't engaged in high-risk behavior like drinking, overeating, or smoking, they had no idea where their health problems had come from. They certainly didn't connect them to the past, because they'd left the past behind. Right?

The truth is that despite all their hard work, people like Evan who have had adverse childhood experiences are still at greater risk for developing chronic illnesses, like cardiovascular disease, and cancer.

But why? How does exposure to stress in childhood crop up as a health problem in middle age or even retirement? Are there effective treatments? What can we do to protect our health and our children's health?

In 2005, when I finished my pediatrics residency at Stanford, I didn't even know to ask these questions. Like everyone else, I had only part of the story. But then, whether by chance or by fate, I caught glimpses of a story yet to be told. It started in exactly the place you might expect to find high levels of adversity: a low-income community of color with few resources, tucked inside a wealthy city with all the resources in the world. In the Bayview Hunters Point neighborhood of San Francisco, I started a community pediatric clinic. Every day I witnessed my tiny patients dealing with overwhelming trauma and stress; as a human being, I was brought to my knees by it. As a scientist and a doctor, I got up off those knees and began asking questions.

My journey gave me, and I hope this book will give you, a radically different perspective on the story of childhood adversity — the whole story, not just the one we think we know. Through these pages, you will better understand how childhood adversity may be playing out in your life or in the life of someone you love, and, more important, you will learn the tools for healing that begins with one person or one community but has the power to transform the health of nations.

I

Discovery

1

Something's Just Not Right

AS I WALKED INTO an exam room at the Bayview Child Health Center to meet my next patient, I couldn't help but smile. My team and I had worked hard to make the clinic as inviting and family-friendly as possible. The room was painted in pastel colors and had a matching checkered floor. Cartoons of baby animals paraded across the wall above the sink and marched toward the door. If you didn't know better, you'd think you were in a pediatric office in the affluent Pacific Heights neighborhood of San Francisco instead of in struggling Bayview, which was exactly the point. We wanted our clinic to be a place where people felt valued.

When I came through the door, Diego's eyes were glued to the baby giraffes. *What a super-cutie,* I thought as he moved his attention to me, flashed me a smile, and checked me out through a mop of shaggy black hair. He was perched on the chair next to his mother, who held his three-year-old sister in her lap. When I asked him to climb onto the exam table, he obediently hopped up and started swinging his legs back and forth. As I opened his chart, I saw his birth date and looked up at him again — Diego was a cutie *and* a shorty.

Quickly I flipped through the chart, looking for some objective data to back up my initial impression. I plotted Diego's height on the growth curve, then I double-checked to be sure I hadn't made a mistake. My newest patient was at the 50th percentile for height for a four-year-old.

Which would have been fine, except that Diego was seven years old.

That's weird, I thought, because otherwise, Diego looked like a totally normal kid. I scooted my chair over to the table and pulled out my stethoscope. As I got closer I could see thickened, dry patches of eczema at the creases of his elbows, and when I listened to his lungs, I heard a distinct wheezing. Diego's school nurse had referred him for evaluation for attention deficit hyperactivity disorder (ADHD), a chronic condition characterized by hyperactivity, inattention, and impulsivity. Whether or not Diego was one of the millions of children affected by ADHD remained to be seen, but already I could see his primary diagnoses would be more along the lines of persistent asthma, eczema, and growth failure.

Diego's mom, Rosa, watched nervously as I examined her son. Her eyes were fixed on Diego and filled with concern; little Selena's gaze was darting around the room as she checked out all the shiny gadgets.

"Do you prefer English *o Español*?" I asked Rosa.

Relief crossed her face and she leaned forward.

After we talked — in Spanish — through the medical history that she had filled out in the waiting room, I asked the same question I always do before jumping into the results of the physical exam: Is there anything specific going on that I should know about?

Concern gathered her forehead like a stitch.

"He's not doing well in school, and the nurse said medicine could help. Is that true? What medicine does he need?"

"When did you notice he'd started having trouble in school?" I asked.

There was a slight pause as her face morphed from tense to tearful.

"*¡Ay, Doctora!*" she said and began the story in a torrent of Spanish.

I put my hand on her arm, and before she could get much further, I poked my head out the door and asked my medical assistant to take Selena and Diego to the waiting room.

The story I heard from Rosa was not a happy one. She spent the next ten minutes telling me about an incident of sexual abuse that had happened to Diego when he was four years old. Rosa and her husband had taken in a tenant to help offset the sky-high San Francisco rent. It was a family friend, someone her husband knew from his work in construction. Rosa noticed that Diego became more clingy and withdrawn after the man arrived, but she had no idea why until she came home one day to find the man in the shower with Diego. While they had immediately kicked the man out and filed a police report, the damage was done. Diego started having trouble in preschool, and as he moved up, he lagged further and further behind academically. Making matters worse, Rosa's husband blamed himself and seemed angry all the time. While he had always drunk more than she liked, after the incident it got a lot worse. She recognized the tension and drinking weren't good for the family but didn't know what she could do about it. From what she told me about her state of mind, I strongly suspected she was suffering from depression.

I assured her that we could help Diego with the asthma and eczema and that I'd look into the ADHD and growth failure. She sighed and seemed at least a little relieved.

We sat in silence for a moment, my mind zooming around. I believed, ever since we'd opened the clinic in 2007, that something medical was happening with my patients that I couldn't quite understand. It started with the glut of ADHD cases that were referred to me. As with Diego's, most of my patients' ADHD symptoms didn't just come out of the blue. They seemed to occur at the highest rates in patients who were struggling with some type of life disruption or trauma, like the twins who were failing classes and getting into fights at school after witnessing an attempted murder in their home or the three brothers whose grades fell precipitously after their parents' divorce turned violently acrimonious, to the point where the family was ordered by the court to do their custody swaps at the Bayview police station. Many patients were already on ADHD medication; some were even on antipsychotics. For a number of patients, the medication seemed to be helping, but for many it clearly wasn't. Most of the time I couldn't make the ADHD diagnosis. The diagnostic criteria for ADHD told me I had to rule out other explanations for ADHD symptoms (such as pervasive developmental disorders, schizophrenia, or other psychotic disorders) before I could diagnose ADHD. But what if there was a more nuanced answer? What if the cause of these symptoms — the poor impulse control, inability to focus, difficulty sitting still — was not a mental disorder, exactly, but a biological process that worked on the brain to disrupt normal functioning? Weren't mental disorders simply biological disorders? Trying to treat these children felt like jamming unmatched puzzle pieces together; the symptoms, causes, and treatments were close, but not close enough to give that satisfying click.

I mentally scrolled back, cataloging all the patients like Diego and the twins that I'd seen over the past year. My mind went immediately to Kayla, a ten-year-old whose asthma was particularly difficult to control. After the last flare-up, I sat down with mom and patient to meticulously review Kayla's medication regimen. When I asked if Kayla's mom could think of any asthma triggers that we hadn't already identified (we had reviewed everything from pet hair to cockroaches to cleaning products), she responded, "Well, her asthma does seem to get worse whenever her dad punches a hole in the wall. Do you think that could be related?"

Kayla and Diego were just two patients, but they had plenty of company. Day after day I saw infants who were listless and had strange rashes. I saw kindergartners whose hair was falling out. Epidemic levels of learning and behavioral problems. Kids just entering middle school had depression. And in unique cases, like Diego's, *kids weren't even growing.* As I recalled their faces, I ran an accompanying mental checklist of disorders, diseases, syndromes, and conditions, the kinds of early setbacks that could send disastrous ripples throughout the lives to come.

If you looked through a certain percentage of my charts, you would see not only a plethora of medical problems but story after story of heart-wrenching trauma. In addition to the blood pressure reading and the body mass index in the chart, if you flipped all the way to the Social History section, you would find parental incarcerations, multiple foster-care placements, suspected physical abuse, documented abuse, and family legacies of mental illness and substance abuse. A week before Diego, I'd seen a six-year-old girl with type 1 diabetes whose dad was high for the third visit in a row. When I asked him about it, he assured me I shouldn't worry because the weed helped to quiet the voices in his head. In the first year of my practice, seeing roughly a thousand patients, I diagnosed not one but *two* kids with autoimmune hepatitis, a rare disorder that typically affects fewer than three children in one hundred thousand. Both cases coincided with significant histories of adversity.

I asked myself again and again: *What's the connection?*

If it had been just a handful of kids with both overwhelming adversity and poor health outcomes, maybe I could have seen it as a coincidence. But Diego's situation was representative of hundreds of kids I had seen over the past year. The phrase *statistical significance* kept echoing through my head. Every day I drove home with a hollow feeling. I was doing my best to care for these kids, but it wasn't nearly enough. There was an underlying sickness in Bayview that I couldn't put my finger on, and with every Diego that I saw, the gnawing in my stomach got worse.

• • •

For a long time the possibility of an actual biological link between childhood adversity and damaged health came to me as a question that lingered for only a moment before it was gone. *I wonder ... What if ... It seems like ...* These questions kept popping up, but part of the problem in putting the pieces together was that they would emerge from situations occurring months or sometimes years apart. Because they didn't fit logically or neatly into my worldview at those discrete moments in time, it was difficult to see the story behind the story. Later it would feel obvious that all of these questions were simply clues pointing to a deeper truth, but like a soap-opera wife whose husband was stepping out with the nanny, I would understand it only in hindsight. It wasn't hotel receipts and whiffs of perfume that clued me in, but there were plenty of tiny signals that eventually led me to the same thought: *How could I not have seen this? It was right in front of me the whole damn time.*

I lived in that state of not-quite-getting-it for years because I was doing my job the way I had been trained to do it. I knew that my gut feeling about this biological connection between adversity and health was just a hunch. As a scientist, I couldn't accept these kinds of associations without some serious evidence. Yes, my patients were experiencing extremely poor health outcomes, but wasn't that endemic to the community they lived in? Both my medical training and my public-health education told me that this was so.

That there is a connection between poor health and poor communities is well documented. We know that it's not just how you live that affects your health, it's also *where* you live. Public-health experts and researchers refer to communities as "hot spots" if poor health outcomes on the whole are found to be extreme in comparison to the statistical norm. The dominant view is that health disparities in populations like Bayview occur because these folks have poor access to health care, poor quality of care, and poor options when it comes to things like healthy, affordable food and safe housing. When I was at Harvard getting my master's degree in public health, I learned that if I wanted to improve people's health, the best thing I could do was find a way to provide accessible and better health care for these communities.

Straight out of my medical residency, I was recruited by the California Pacific Medical Center (CPMC) in the Laurel Heights area of San Francisco to do my dream job: create programs specifically targeted to address health disparities in the city. The hospital's CEO, Dr. Martin Brotman, personally sat me down to reinforce his commitment to that. My second week on the job, my boss came into my office and handed me a 147-page document, the *2004 Community Health Assessment* for San Francisco. Then he promptly went on vacation, giving me very little direction and leaving me to my own ambitious devices (in hindsight, this was either genius or crazy on his part). I did what any good public-health nerd would do — I looked at the numbers and tried to assess the situation. I had heard that Bayview Hunters Point in San Francisco, where much of San Francisco's African American population lived, was a vulnerable community, but when I looked at the 2004 assessment, I was floored. One way the report grouped people was by their zip code. The leading cause of early death in seventeen out of twenty-one zip codes in San Francisco was ischemic heart disease, which is the number-one killer in the United States. In three zip codes it was HIV/AIDS. But Bayview Hunters Point was the only zip code where the number one cause of early death was violence. Right next to Bayview (94124) in the table was the zip code for the Marina district (94123), one of the city's more affluent neighborhoods. As I ran my finger down the rows of numbers, my jaw dropped. What they showed me was that if you were a parent raising your baby in the Bayview zip code, your child was two and a half times as likely to develop pneumonia than a child in the Marina district. Your child was also six times as likely to develop asthma. And once that baby grew up, he or she was *twelve* times as likely to develop uncontrolled diabetes.

I had been hired by CPMC to address disparities. And, boy, now I saw why.

. . .

Looking back, I think it was probably a combination of naïveté and youthful enthusiasm that spurred me to spend the two weeks that my boss was gone drawing up a business plan for a clinic in the heart of the community with the greatest need. I wanted to bring services to the people of Bayview rather than asking them to come to us. Luckily, when my boss and I gave the plan to Dr. Brotman, he didn't fire me for excessive idealism. Instead, he helped me make the clinic a reality, which still kind of blows my mind.

The numbers in that report had given me a good idea of what the people of Bayview were up against, but it wasn't until March of 2007, when we opened the doors to CPMC's Bayview Child Health Center, that I saw the full shape of it. To say that life in Bayview isn't easy would be an understatement. It's one of the few places in San Francisco where drug deals happen in plain sight of kindergartners on their way to school and where grandmas sometimes sleep in bathtubs because they're afraid of stray bullets coming through the walls. It's always been a rough place and not only because of violence. In the 1960s, the U.S. Navy decontaminated radioactive boats in the shipyard, and up until the early 2000s, the toxic byproducts from a nearby power plant were routinely dumped in the area. In a documentary about the racial strife and marginalization of the neighborhood, writer and social critic James Baldwin said, "This is the San Francisco that America pretends does not exist."

My day-to-day experience working in Bayview tells me that the struggles are real and ever present, but it also tells me that's not the whole story. Bayview is the oily concrete you skin your knee on, but it's also the flower growing up between the cracks. Every day I see families and communities that lovingly support each other through some of the toughest experiences imaginable. I see beautiful kids and doting parents. They struggle and they laugh and then they struggle some more. But no matter how hard parents work for their kids, the lack of resources in the community is crushing. Before we opened the Bayview Child Health Center, there was only one pediatrician in practice for over ten thousand children. These kids face serious medical and emotional problems. So do their parents. And their grandparents. In many cases, the kids fare better because they are eligible for government-assisted health insurance. Poverty, violence, substance abuse, and crime have created a multigenerational legacy of ill health and frustration. But still, I believed we could make a difference. I opened my practice there because I wasn't okay with pretending the people of Bayview didn't exist.

. . .

Patients like Diego and Kayla were exactly why I came to Bayview. For as long as I could remember, I knew this was the problem I wanted to focus on, the type of community I wanted to serve. I had gotten the best medical education I could, earned a master's in public health, and was well trained in how to work with vulnerable communities to improve access to health care. After years of schooling, I had faith in the dominant academic view: if you improve people's access to quality health care, you will move the needle toward better health. I knew what boxes to check and I was ready to go. When I first got to Bayview, I thought all I had to do was put it in motion — start giving people great care, make it easy for them to get it affordably, and watch that needle move toward healthier kids. It seemed simple enough.

There was some pretty basic care that we could quickly implement, and by employing standardized clinical protocols, our clinic was able to dramatically improve outcomes on some things, like increasing immunization rates and decreasing asthma hospitalizations. So I was feeling pretty good for a while. But then, as I was handing out vaccines and inhalers, I started to wonder: If we were doing everything right, why didn't we see any indication that we could make a dent in this community's dramatically reduced life expectancy? My patients kept coming back with high rates of illnesses, and I had the sinking feeling that when they grew up, their kids would keep coming back too. Despite the checked boxes, despite the great care, and despite more health-care access than the community had seen in a generation — the needle in Bayview only bounced.

· · ·

After my medical assistant had taken Diego and his sister into the waiting room and Rosa had told me some of his history, the two of us sat momentarily with our thoughts. I could only imagine the guilt, worry, and hope swimming around in her head. Regardless of our individual thought soups, both of our faces cracked into helpless smiles when Diego slid through the door, cross-eyed and goofy. Rosa stood up and I took note of her size. She was a stout woman, but height-wise, she wasn't below the range of normal. Diego, however, was so small that he did not even come close to the growth curve for a seven-year-old boy. I remember mentally clicking through the protocol for evaluation and treatment of growth failure. Which makes sense; that's what doctors do. You see a problem — abnormal development or disease — and you try to right the ship. But this time a simple question surfaced: *What am I missing?*

· · ·

There is a widely known parable that students all learn on day one in public-health school, and it happens to be based on a true story. In late August 1854, there was a severe cholera outbreak in London. The Broad Street area in Soho was the epicenter, with a hundred and twenty-seven dead in the first three days and more than five hundred dead by the second week of September. Back then the dominant theory was that diseases like cholera and bubonic plague were spread through unhealthy air. John Snow, a London physician, was skeptical of this "miasma theory" of disease. By canvassing the residents of the Broad Street neighborhood, he was able to deduce the pattern of the disease. Incidences were all clustered around a water source: a public well with a hand pump. When Snow convinced local officials to disable the well by removing the pump's handle, the outbreak subsided. At the time, no one wanted to accept Snow's hypothesis that the disease was spread not through the air but by the more unpleasant fecal-oral route, but a few decades later, science would catch up to him, and the miasma theory would be replaced by germ theory.

As budding public-health crusaders, my classmates and I focused on the sexy part of the parable of the well, the bit where Snow topples the miasma theory. But I also took away a larger lesson: If one hundred people all drink from the same well and ninety-eight of them develop diarrhea, I can write prescription after prescription for antibiotics, or I can stop and ask, "What the hell is in this well?"

I had been about to walk past the well to do the standard evaluation for Diego's growth failure, but this time something made me think about the case in front of me a little differently. Maybe it was the extreme presentation. Maybe I had finally seen enough cases to start putting the pieces together. Whatever the reason, I couldn't get away from the nagging feeling that Diego's terrible trauma and his health problems weren't just a coincidence.

But before I could look into the well for the answer to Diego's, or any of my patients', problems, I needed a few more data points. The first step in Diego's case was to order a bone-age study, an x-ray of the left wrist that can be used to determine a child's skeletal maturation based on the size and shape of the bones. After drawing some labs and requesting his growth charts from the clinic where he had previously been seen, I handed Rosa the order form for the x-ray and sent my newest patient on his way.

Days later, I received the report from the radiologist. It confirmed that Diego's skeletal maturity was consistent with that of a four-year-old. But Diego's labs didn't show low levels of growth hormone or any other hormone that might account for why he wasn't growing. I had some important data in front of me: The trauma had happened at age four and he had gained very little vertical height since then. He also had the bone age of a four-year-old. But by all accounts, Diego wasn't malnourished and didn't have any evidence of a hormonal disorder. There didn't seem to be a readily available medical explanation for Diego's stature.

My next call was to Dr. Suruchi Bhatia, a pediatric endocrinologist at California Pacific Medical Center. I sent her the x-ray report and Diego's labs and asked whether she thought the sexual assault of a four-year-old could lead to that child's growth arrest.

"Is that even something you've seen before?" I asked, finally verbalizing what had been bugging me all week.

"The not-so-simple answer? Yes."

Oh, man, I thought. *Now I really have to find out what the hell is going on.*

. . .

I couldn't stop thinking about how extreme this physical presentation was. If what was in the "well" in Bayview was adversity, Diego had experienced a high dose of it, the equivalent of drinking a jug of cholera-infested water. If I could figure out what was going on with Diego on a biochemical level, maybe I would learn what was going on with *all* of my patients. Maybe it was even the key to what was going on in the community at large. I had four major questions to answer: Was the exposure (trauma/adversity) at the bottom of the well making people sick? How? Could I prove it? And most important, what could I do about it medically?

One immediate problem with getting to the bottom of this larger connection between adversity and ill health was that at times, there was an overwhelming number of factors to consider — my patients' different upbringings, their genetic histories, their environmental exposures, and, of course, their individual traumas. I already knew it wasn't going to be as simple as identifying a shared water source and a single type of bacterium. With Diego, an incident of abuse had acted as a catalyst that (presumably) set off a biochemical chain reaction resulting in growth arrest. But all kinds of wild things had to go on, and *keep* going on, hormonally and cellularly, for the body to react in such an extreme way. Figuring this out would take some doing. I saw the next months of my life flash in front of me; it was nothing but PubMed, granola bars, and eye strain.

That day at the clinic, I stayed well into the evening, combing through patient charts for patterns I might have missed. Eventually I got up and began to pace. All the patients and staff had gone home, so I was free to wander without distraction. I meandered through the waiting room, stopping to smile at the mini-furniture and the primary-color footprints stenciled on the rug. These things reminded me yet again that my patients were normal kids, regardless of what they had been through or would go through.

When I was first working for CPMC in Laurel Heights, my favorite part of the job was examining newborns. Years later, I did identical exams on the newborns of Bayview, and I found that their little hearts sounded the same under my stethoscope. When I put a gloved finger in an infant's mouth, the same adorable suckling reflex kicked in. They all had the same soft spots on the tops of their heads where the skull bones hadn't quite closed yet. These babies came into the world no different than the ones born in Laurel Heights, yet as I did newborn exams in Bayview, I knew that these human beings' lives would, according to the statistics, be twelve years shorter than the lives of the children in Laurel Heights. Not because their hearts were made differently or because their kidneys didn't function the same way, but because somewhere in the future, something in their bodies would change — something that would alter the trajectory of their health for the rest of their lives. At the beginning, they are equal, these beautiful bundles of potential, and knowing that they won't always be is enough to break your heart.

· · ·

I walked into the exam room just before leaving for home, flipped on the light, and looked at the animals stenciled on the wall — lions, giraffes, horses, and, strangely, a single, solitary frog. My gaze lingered there. Maybe it was that the frog was oddly solo, or maybe it was just the brain's mysterious way of connecting the dots, but suddenly I remembered the Hayes lab at the University of California, Berkeley. When I was twenty years old I logged some serious hours there, and frogs were a big part of it. The Hayes lab was an amphibian research lab where the inimitable Dr. Tyrone Hayes was studying the effects of corticosteroids (stress hormones) on tadpoles at different stages of their development. The ghosts of research past flooded my brain, intersecting with the problem I'd been fighting all day: Everything I'd learned in my training told me that adversity was a social determinant of poor health outcomes, but what was never examined was *how* it affected physiology or biological mechanisms. There wasn't any research that I could fall back on to help me understand how my patients' traumatic experiences could be affecting their biology and their health.

Or maybe there was.

Maybe to figure out what was going on with Diego and all the little tadpoles in Bayview, I had to look for clues in more cold-blooded circles.

2

To Go Forward, Go Back

IF IT'S TRUE THAT parents are a child's first teacher, the fact that my dad was a professor of biochemistry who had a penchant for instructive chaos probably says a lot about me. At one point in the eighties, my parents were raising five kids under the age of ten, so we probably left them little choice but to get creative on the parenting front. My father, Dr. Basil Burke, is a Jamaican immigrant, and if I can dad-brag for a second, when the Institute of Jamaica gave out the Centennial Medal to honor its hundredth anniversary, Bob Marley got one for music and my dad got one for chemistry. To this day when he babysits my kids, I never know what I'm going to come home to. A mysterious chalky white substance coating every inch of the stove? A carefully deconstructed water filter? Three raw shrimp on the counter next to three cooked shrimp? It's always a surprise with my pops.

I knew from an early age that he wasn't like other dads. As a biochemist, he turned every one of our kid "experiments" into an opportunity (ahem, demand) for discovery. When he came home from work to find me and my four brothers lobbing sharp-nosed paper airplanes at one another with wild glee, he didn't yell at us to stop before we poked someone's eye out. Instead, he sprang into action, commanding us to take measurements on the floor and time our throws. If you calculated how long it took an airplane to get from point A to point B, you could determine its velocity. And from that, knowing that gravity caused an object to accelerate at 9.8 meters per second squared, you could determine the lift under the wings and extrapolate the best angle at which to release the plane in order to hit someone. In hindsight, I see that this kind of intervention was actually brilliant parenting, because inevitably my brothers would groan, drop their weapons, and get the hell out of there. I, however, couldn't get enough. My dad brought physics, chemistry, and biology to bear on everything from curdled milk in the fridge to the curry stain on my blouse that mysteriously turned from yellow to purple the minute I touched it with a bar of soap. While my mother was none too pleased about the stench of sour milk or a ruined blouse, I learned something that became fundamental to my adult worldview: there is a molecular mechanism behind every natural phenomenon — you just have to look for it.

A decade later, during my internship in the Hayes lab, I realized that a big part of what made my dad a great scientist was the intense joy he took in the process. I'd come to understand that doing science as a professional was not the same as blowing stuff up as a kid. There was a whole lot of mind-numbing pipetting and data entry, so it was easy to miss the forest for the trees. But the best scientists didn't. They used their excitement and enthusiasm as a bridge from the mundane to the revelatory. If you approach your experiments simply as plug and play — either they work or they don't — then you're missing the potential for a happy accident. Day to day, good scientists actively engineer the conditions for discovery by making the most of accidents. Like my curry-stained blouse, a botched experiment can be a gateway to an unexpected truth. As a kid, I saw how this worked by watching my dad. As a college student, I learned it at the hands of Dr. Tyrone B. Hayes.

Dr. Hayes was the antithesis of the typical Berkeley science professor. Just twenty-seven years old at the time I worked under him, he was one of the youngest professors on the science faculty. Not only was he brilliant, he was my only African American science professor at Cal, and he happened to have a wicked sense of humor and an eloquently foul mouth. No one even called him Dr. Hayes; he was just straight-up Tyrone. Thanks to him, ours was by far the coolest lab in the building.

. . .

The Hayes lab specialized in groundbreaking amphibian endocrine research, so naturally, tadpoles and toads were my life for every spare hour of my senior year at Berkeley. The research I was working on would turn out to be one of Hayes's most important accidents. Hayes's experiment started with a hypothesis about sexual development in toads and was designed to figure out the impact of different kinds of steroid hormones (testosterone, estrogen, corticosterone) on gonad differentiation—basically, whether tadpoles would develop into female or male adult toads. Hormones are an organism's chemical messengers; the information they carry through the bloodstream stimulates a wide range of biological processes. He exposed the tadpoles to a range of steroids over different periods of their development and to his surprise found there was no effect on the gonads. A whole lot of time and thought went into these experiments, yet in the end, no measurable difference was observed. A bummer, to say the least. But while I was triple-checking tissue samples under the microscope, Hayes was training a creative eye on the disappointing results. What he found was that while none of the steroids had an impact on the *sexual* development of the tadpoles, some of the steroids had an effect on their growth and subsequent metamorphosis. The most eyebrow-raising effects were observed when Hayes exposed tadpoles to corticosterone.

For Hayes, the impact this hormone had on the growth of tadpoles was interesting enough for him to think about throwing his experimental darts in a totally different direction. Corticosterone is a stress hormone—its equivalent in humans is cortisol—so Hayes put on his frog suit and tried to imagine a stressful scenario for a tadpole. What he came up with was simple enough: a pond starts drying up and suddenly there are too many tadpoles and not enough water. He hypothesized that a stress response in that situation could be adaptive, meaning that when the tadpole got stressed by all the other pushy tadpoles and the receding water, its glands would release corticosterone, which would jump-start the process of metamorphosis and turn its tail into legs. Now the newly minted toad could jump out of the pond and leave all the other tadpole-chumps behind. Bingo! Adaptation.

That was the idea, at least. Turns out Hayes was mostly right, but as always, *how he was wrong* was where things get interesting. If the toads-to-be were exposed to corticosterone late in development, it *did* speed up metamorphosis, allowing for the adaptive, timely leap out of the pond. But if the toads were exposed to the steroid early in development, it actually *inhibited* their growth. And it had other unexpected negative effects, such as decreasing immune function, diminishing lung function, causing osmoregulatory problems (high blood pressure), and impairing neurological development. If the tadpoles were exposed to corticosterone for a prolonged period, the same problems occurred. The tadpoles' stress response to overcrowding was adaptive, but *only* if it happened at the right time during development.

Why was exposure to the stress hormone so bad for the younger tadpoles? That's the tricky part. High levels of corticosterone affects the function of other hormones and body systems. For the tadpoles, early and prolonged exposure to corticosterone threw all of these other hormone levels and biological processes out of whack. The effects were *maladaptive,* meaning that instead of helping the tadpole thrive and survive, the response made things much, much worse. In fact, early exposure often led not only to irreversible developmental changes but, eventually, to death. For instance, levels of corticosterone can have an impact on levels of thyroid hormone, which regulates metabolism. In the case of the tadpoles, corticosterone knocked out the thyroid hormone completely, which is why those tadpoles didn't grow and develop to the metamorphosis stage. Corticosterone also affects the production of surfactant, which plays a key role in lung development, allowing them to absorb oxygen out of the air.

Because I was on the premed track, I had learned in anatomy and physiology how hormones work together in a kind of symphony of homeostasis (the body's biological balance or equilibrium). But it wasn't until I worked in the Hayes lab that I really *got* it. The unlucky frogs served as a critical object lesson. If you have the right amount of each hormone, they all work together to keep the body functioning normally, but if you change one of those levels, the delicate interplay gets thrown off. This kind of hormonal imbalance can have direct and indirect effects. For instance, an increase in corticosteroids can directly affect blood pressure, but it can also indirectly affect growth and development by altering how other hormones do *their* jobs. How hormones affect one another and, as a result, the human body can be complicated, but it's hugely important.

Another eye-opener for me at the Hayes lab was the compulsory evolutionary stress-response primer that everyone got on the first day of work. It's easy (kind of) to memorize the impacts of various hormone interactions in the body: if A and B, then C. Science in school is a never-ending pageant of flow charts, infographics, formulas, and calculations; the *what* of the human body, if you will. Looking at biology from the evolutionary perspective, as Hayes's tadpoles taught us to do, we got something just as important: the *why*. Most of us came in with an understanding of the biological cause and effect of physiological processes in the ideal, adaptive state; we left with a fascination for decoding the cause and effect in a far-from-ideal, maladaptive state.

For most of early human history, the biggest stressors (stress-inducing events) were predators (short-term stressors) and food shortages (long-term stressors). Back in the day on the savanna, a major purpose of cortisol was to help the body manage that long-term stress. Maintaining homeostasis is the key to survival, so cortisol shows up when the body detects a change in the environment that threatens to push it off balance. With a shortage of supermarkets (and iPhone apps) in prehistoric Africa, early humans spent most of their days finding food, killing food, and preparing food for eating. When times were tough, the body would detect a lack of nutrients and begin the chain reaction that is the stress response.

One of the biggest parts of this process is the increased production of cortisol, and a major effect of cortisol is an increase in blood sugar. The brain needs enough blood sugar to be able to think and plan, so this extra punch of cortisol helps keep blood sugar on an even keel, despite a shortage of gazelle BBQ. The steady stream of glucose swimming in your veins also helps fuel your muscles, so in the event that you *do* see a gazelle, you have the energy to chase it down. Cortisol also helps maintain normal blood pressure by regulating the body's water and salt levels. And it inhibits growth and reproduction, because if you are living through a food crisis, it's not a good time to be doing any optimistic long-term family planning; it makes more sense to divert all available energy to the problem at hand. Cortisol has these effects and more, and not just when there is a lack of food but also when there is a physical threat (lions, for instance), an injury, or an environmental stressor (earthquake!). Every time the stress response gets triggered, the same basic biological processes kick in. The difference between the ancient adult human surviving a bad hunting season and the tadpole getting a lethal dose of stress is the *timing and duration* of the exposure to the stress hormone. In the case of the hunter, the process was adaptive (good for survival) because it happened in adulthood; in the case of the tadpole, the process was maladaptive (bad for survival) because it happened in childhood (or tadpole-hood), too early in development.

. . .

In the days following my initial visit with Diego, I thought about the Hayes lab a lot — what I'd learned about the stress response, what I'd learned about development, and what I'd learned about how to creatively approach a problem. The last part was what stayed with me as I reviewed Hayes's old research paper on corticosterone and its role in metamorphosis. But while the tadpoles and toads gave me a solid appreciation for how stress hormones can affect development, I recognized that these were animal studies. The tadpoles were given significant doses of corticosterone, and as a result, the effects were dramatic. That made sense, but like a lot of animal studies, there was no guarantee that it would translate directly to humans. And no one tried the experiment on humans owing to the small ethical problem of giving people massive doses of stress hormones. So there were no studies evaluating the effects of massive doses of stress hormones on humans, let alone on children. Or were there?

. . .

I was a third-year resident working in the pediatric intensive care unit (PICU) at Lucile Packard Children's Hospital at Stanford, and Sarah P. was a beautiful six-year-old girl who woke up one morning paralyzed from the waist down. After an extensive workup, we finally determined the cause: ADEM, acute disseminated encephalomyelitis.

ADEM is a rare autoimmune disease in which the body's immune system attacks myelin, the insulating sheaths that surround nerve fibers and allow nerve impulses to travel quickly through the body. Sarah's parents were understandably terrified. The treatment for ADEM is high doses of the steroid prednisone, which is essentially a synthetic version of cortisol. The hope is that the "stress dose" of steroids will suppress the misguided immune system so that nerve function can recover. As I was writing the order for the prednisone, my supervising physician reminded me to also write what doctors call standing orders. This is an automatic protocol put in place for every time a particular medication is given. For stress doses of steroids, standing orders addressed what to do if Sarah P. had any of the predictable side effects. In the pediatric ICU, decades of experience have shown physicians that most patients who get high doses of prednisone have the same types of problems. So the standing order goes something like this: (1) If blood pressure reaches [X], give [Y] blood pressure medication; (2) if blood sugar goes above [X], start insulin drip at [Y] rate; (3) if the patient becomes psychotic and tries to rip out her IV, give [X] antipsychotic at [Y] dosage.

When I got to this particular part of memory lane, I couldn't help yelling, *"Rhaatid!"* (Jamaican patois for "Oh my goodness!"). I realized that the effects of a stress dose of steroids on a child were not only known but codified in the hospital's protocols for care. Medical protocols are put in place when the side effects of a certain medication are *so predictable* that it's worth setting up a system for addressing them. This is one of those unique scenarios where clinical experience becomes living research. The doctors at Stanford observed the side effects that patients who received stress doses of steroids exhibited, then they investigated what they thought was going on and made adjustments to care until they found the best way to treat those side effects. It may not be ethical to run a premeditated independent experiment to test how human children react to stress hormones, but observing their reactions during a course of lifesaving treatment certainly is. Over time, the successful interventions the doctors implemented became the clinical guidelines for managing the side effects of prednisone. Sarah P.

was the lucky beneficiary of this, getting enough of the medicine for her to improve (and recover, I'm thankful to say), but not so much that it created larger problems for her.

Suddenly, my patients' physical reactions didn't seem so crazy. If their systems were flooded with stress hormones just like Sarah's or the tadpoles', it stood to reason that their bodies, including their blood pressure, blood sugar, and neurological functions, might react in similar ways; all could be seen as side effects of stress hormones. It made biological sense that a high dose of stress hormones at the wrong developmental stage could have an outsize impact on my patients' downstream health. It was exactly what happened to the younger tadpoles versus the closer-to-metamorphosis froglets — the difference between adaptive and maladaptive reactions is all about the *when*.

An extreme example of the impact of timing when it comes to hormones is a condition called hypothyroidism. Many of us know someone or have heard of someone who has an underactive thyroid. It basically means that the thyroid gland is not producing enough thyroid hormone, so a person's metabolism slows down and he or she develops dry skin, brittle hair, and the most well-known symptom: weight gain. While somewhere around ten million adults have this condition, it often takes a long time to diagnose it. But the good news is that symptoms in adults tend to be relatively minor, and the treatment is readily available.

When hypothyroidism occurs in infancy, though, it's a whole different ball game. The condition, once cruelly called cretinism, can result in significantly impaired physical and mental growth. Generations of children suffered severe symptoms because physicians caught the disorder too late, but now newborns are screened for hypothyroidism. When identified early, the condition is easily treated with thyroid hormone, which means cretinism is now extremely rare in the developed world. But it's still a great example of just how critical timing is: a lack of thyroid hormone in the body has wildly different effects depending on when it happens. In adulthood, it's minor and treatable. In childhood, it's profound.

· · ·

When it came to Diego, the timing of his symptoms worried me. I feared that the dose of stress he had experienced was high enough to overload his system and that it might be the underlying cause of his symptoms. The same went for my other patients.

But what about the rest of the community? Plenty of the adult population had experienced adversity and trauma on par with Diego's during their own childhoods. Because my patients were kids, I found out about the trauma they had experienced through their parents or caregivers. Often the parents had experienced far more adversity than the children they were bringing to the clinic; the moms, dads, aunties, and uncles I came to know over the years periodically shared their own histories of being physically, verbally, or sexually abused, growing up with domestic violence, or even witnessing someone get stabbed or shot. Now they had arthritis, failing kidneys, heart disease, chronic lung disease, and cancer. Most had grown up in Bayview or similar communities, and I couldn't help but wonder about the long-term effects these early exposures had had on the health of entire generations.

There was no question that the people of Bayview, my patients included, experienced higher doses of stress than your average American. I thought about little Sarah P. and the standing orders for steroid side effects. If the adults of Bayview were once children who experienced stress doses of hormones during the critical stages of development, what were the side effects that we could see as a result?

The answer was right there in the 2004 *Community Health Assessment* I'd read on my first day of work.

There are thousands of Bayviews all over the United States, not to mention the globe. In public-health school, I heard lectures about the extent of health disparities between vulnerable communities (such as those with high percentages of low-income, recent immigrants, or communities of color) and wealthier neighborhoods, and as a black woman from an immigrant family in America, I felt like I was being told that water was wet. What I was looking for was the *why*. I had a distinct memory of sitting in Professor Ichiro Kawachi's classroom in Boston as he presented some striking data about obesity rates in high-risk communities and asking myself, *Could this be related to cortisol? Is it possible that the daily threat of violence and homelessness breath-*

*ing down your neck is not only associated with poor health but poten-
tially the cause of it?* It occurred to me, horribly, that people living in
crowded public housing in Chicago might not be all that different
from tadpoles living in a shrinking pond.

But now, sitting in Bayview, I saw that what people experienced in
childhood could be enough to set them on a devastating medical trajectory. The very idea that the events of childhood could affect people's
health for the rest of their lives was scary, but if the stress-response system was indeed the mechanism in play, it opened up a huge runway
for change. It meant that if we figured out the problem early enough
in a child's development, we could make a significant impact on his or
her later life. Length of exposure and timing were both critical when
it came to the effects of corticosterone on the tadpoles. With the kids
of the Stanford PICU, we knew there were measures we could take to
address the side effects of stress hormones before they caused a problem. Could my colleagues and I create a standing order for patients
like Diego? What would that look like? I didn't know, but the thought
was enough to fill me with the same kind of electricity that I felt as a
kid when I was working on a problem with my dad and could sense I
was on the right track.

3

Forty Pounds

THE BEAUTY AND THE challenge of working in a clinic like mine was that, regardless of your own needs (sleep!) or desires (lunch!), there was an undertow of urgency that always pulled you back to your patients. After work I sometimes had the luxury of time to investigate the connection between adversity and health, but when I was in the clinic, I had a stack of charts and a waiting room full of sick kids. With Diego in particular, I felt that familiar tug. While I had written him prescriptions for an inhaler and eczema medication, I still needed to tackle the growth arrest. I enlisted the help of Dr. Bhatia once again. I wondered if a course of hormonal therapy might be necessary, but she reminded me that Diego's labs hadn't shown hormonal imbalances, or at least none we could measure. Her experience was that, in cases such as these, medication likely wouldn't help. To my surprise, she said the most effective type of treatment for Diego was *talk* therapy.

Luckily, I already had someone to turn to. The Bayview Child Health Center had received a small grant for patient-support services, and when it came to figuring out what to do with it, I knew just who to ask — the community itself. I understood from my training that building relationships in underserved communities is important for improving health outcomes, which is why I made it a part of my work to help schools and churches plan health fairs, nutrition programs, and asthma-prevention classes. Folks got used to seeing my face in the neighborhood. Many well-meaning people had come and gone in Bayview, leaving a multitude of unfulfilled promises in their wake, but the community was beginning to believe me when I said I was committed to improving the health of their children.

When the grant money for patient support arrived, the answer regarding how to spend it was clear: mental-health services. Though at the time it was pretty unusual for a pediatric office to have a therapist on staff, my colleagues and I knew enough to give the members of the community what they said they needed, not what *we thought* they needed.

But I was nervous about finding the right person to fill the therapist position. We were a nonprofit health center in the middle of Bayview Hunters Point with minimal staff and budget and plenty of intense, unpaid overtime to go around. While that kind of work might have been *my* idea of a dream job, I wasn't crazy enough to think it was everybody's. When Dr. Whitney Clarke walked into my office for an interview, my hopes fell. Even though I certainly knew enough not to judge someone by outward appearances, I still thought, *There is no way this is the guy.*

It would be an understatement to say that someone who looks like Dr. Clarke is not the first image to come to mind when you think of a therapist working in a community like Bayview. He's male, he's white, and he's a dead ringer for Chris Pine (the actor who plays a young Captain Kirk in the new Star Trek films). Basically, he's a walking Abercrombie and Fitch ad. Which to me meant that patients would have trouble trusting him and connecting with him — something of a problem for a therapist in a marginalized, high-needs community. But after we talked for a long time, my initial skepticism started to thaw and I saw something in him that I had a hunch my patients would respond to.

Most of my patients, predictably, pushed back when I referred them to Dr. Clarke. "I'm not taking my child to a white therapist" was a common and understandable refrain. These families were in a vulnerable place, and many had experienced the kind of institutionalized racism that breeds a deep mistrust of outsiders and a reflexive defensiveness. Luckily, by then I had built a strong enough relationship with the community that when I vouched for Dr. Clarke and said

I thought he could make a huge difference for their kids, they trusted me. It was never long before they saw him for who he was: a fiercely caring, easy-to-talk-to skilled practitioner who quickly became a sort of haven for them. I always loved it when these patients' families saw me again months later glowing with a kind of pride about him. Soon, they were vouching for him too.

. . .

After talking with Dr. Bhatia about Diego, I brought Dr. Clarke up to speed and asked what type of therapy plan we should recommend for him. Soon we had connected Rosa with a Spanish-speaking therapist who was experienced in trauma-focused cognitive behavioral therapy (TF-CBT for short), a clinical protocol designed to address the impact of trauma on a child's development by working with both parent and child.

With that knocked off my endless to-do list, I felt better, but although Diego was now on the best treatment plan we could devise, I was still frustrated. I was seeing more and more clearly in my patients a connection between adversity and poor health, but I felt totally unprepared to deal with it. While I was grateful for Dr. Bhatia's guidance regarding Diego's growth, there were many other times when I had no one to call. The previous decade of experiences had led me to trust that what I was seeing was real, but if it was true, why hadn't I learned how to treat it in medical school or residency? Where were the clinical protocols? Where were committees' recommendations to doctors about what to do about this?

Whitney Clarke was often a sounding board for my frustration. Time and again we talked about my hypothesis that adversity was at the root of both the mental-health symptoms he was treating and my most vexing medical cases. Despite his lack of background in endocrinology, it made perfect sense to him. He even reminded me of a few other extreme cases that we'd come up against that fit the Diego stress-symptom mold.

. . .

A couple of months later, Dr. Clarke came to my office and handed me a research paper with a big smile on his face.

"Have you seen this?" he asked.

It was a 1998 article in the *American Journal of Preventative Medicine:* "Relationship of Childhood Abuse and Household Dysfunction to Many of the Leading Causes of Death in Adults: the Adverse Childhood Experiences (ACE) Study," by Dr. Vincent Felitti, Dr. Robert Anda, and colleagues.

"No," I said, sensing by his tone that this was something important.

"You might want to take a break from charting," he said.

"Is this what I *think* it is?"

"Just take a look and then come talk to me," he said.

Before he could even shut the door I was halfway through the abstract. I was only partway through the first page when I experienced a jolt of recognition.

Here it was.

The final puzzle piece that pulled all the others into place.

Everything I had experienced in the past ten years, all of those questions and observations that I couldn't quite put together, suddenly had a linchpin. With my heart knocking in my chest, I started to read aloud the particularly mind-blowing parts of the study, occasionally stopping to whisper-shout in Jamaican patois. The first thing that struck me about Felitti and Anda's research was how incredibly robust it was: they reported data from 17,421 people, which was a large enough number to provide the validation I'd never thought I'd find.

When I finished reading the study, my excitement hadn't diminished. I felt like Neo at the end of the movie *The Matrix* when suddenly the world was dripping with green numbers. Not only was I seeing the full reality of what was all around me, but I *understood it.* According to the ACE Study, I wasn't the only one making connections between the stress of childhood adversity and bad health outcomes. This piece of the puzzle, the final piece of code in the Matrix, was just what I needed to make sense of what was going on with my patients and, more important, to treat them. At the time, I knew that this moment, this understanding, was going to profoundly change my practice, but I had no idea how much it would change my life.

• • •

It was 1985 at the Kaiser obesity clinic in San Diego, and Dr. Vincent Felitti was interviewing his first patient of the day. If you were to stand behind Dr. Felitti in line for soup at the hospital cafeteria or glide past him in the hallway, you would probably be struck by his bearing. *Stately. Composed.* These are the words you might use. Every bit the poised intellectual with a full head of thick, white hair, he looked ready to host the news hour on public television or calmly moderate a debate between acrimonious politicians. He spoke with confidence and authority and was extremely articulate. Which was why when he told me this story, I was blown away to discover that his biggest medical breakthrough had happened because of a verbal slip.

Donna was a fifty-three-year-old woman with debilitating diabetes and a significant weight problem. In a new weight-loss program, she had successfully lost upwards of one hundred pounds two years before, but in the past six months, she'd put it all back on. Felitti felt a conflicting sense of frustration and responsibility. The truth was he didn't really know why Donna had gone off the rails. She had been doing so well and then, after all her hard work and success, she was right back where she started.

Felitti was determined to get to the bottom of it.

He rattled off a list of his usual preliminary questions: How much did you weigh when you were born? How much did you weigh when you started first grade? How much did you weigh when you entered high school? How old were you when you first became sexually active?

But this time, he misspoke.

Instead of asking, "How old were you when you first became sexually active?" he asked, "How much did you weigh when you first became sexually active?"

"Forty pounds," said Donna.

Her answer stopped him short. *Wait a minute, forty pounds?*

He was pretty sure he'd heard her wrong, and for a minute he didn't say anything, but then something made him ask the question again the same way. Maybe she had meant one hundred and forty pounds.

"Sorry, Donna, how much did you weigh when you first became sexually active?"

She went quiet.

He waited for her to speak, sensing there was something here. Working with patients for over two decades had taught him that on the other side of a pregnant pause was usually the diagnostic gold.

"I was forty pounds," Donna said, looking down.

Felitti waited, stunned.

"It was when I was four years old, with my father," she said.

Felitti told me that in the moment, he was shocked, but he struggled not to show his emotions (I knew the feeling all too well). In twenty-three years of working with patients, he had never heard someone tell a story of sexual abuse during a checkup. Nowadays, that would be hard to believe. I wondered if it was because he had never asked or because it was the eighties, when stories of abuse were even more buried than they are today. When I asked him about it, Felitti said he thought he'd probably never asked; he was a doctor, after all, not a therapist.

. . .

Weeks after speaking to Donna, Felitti interviewed another noncompliant patient who was part of the same weight-loss program. Patty had actually started out as a model patient; in a jaw-dropping fifty-one weeks, she had gone from 408 to 132 pounds. Patty and Donna weren't alone. Many other patients were also experiencing great results, some losing up to three hundred pounds in one year on the regimen. Felitti was excited by the outcomes, but the high dropout rate was puzzling. If it had been patients who were still early in the process, the attrition would have been understandable. After all, the fasting regimen they committed to was challenging. But the strange part was that the dropout rate was highest among the most *successful* patients — the very ones who had stuck with it the longest and seen the best results. Just as they were reaching their ideal weights, when they should have been celebrating their hard-won goals, these successful patients suddenly disappeared. They would drop out of the program permanently or leave and come back months later, having regained a majority of the weight they had lost. Felitti and his colleagues were left scratching their heads. They had found what seemed to be a solution for a notoriously intractable problem, yet it was proving unsustainable for no discernible reason.

Felitti was meeting with Patty to try to understand what was going on. He could tell she was on the verge of dropping out of the program because in the past three weeks she had regained thirty-seven pounds. She was going the wrong way, fast. He hoped he could get her back on track before it was too late.

He performed a physical examination on Patty to see if he could determine what was behind the sudden weight gain. Was her heart failing, causing her to retain large amounts of fluid? As far as Dr. Felitti could tell, she wasn't exhibiting the bloating or puffiness that indicated fluid retention associated with heart failure. Was her thyroid out of whack? He took a closer look at her hair, skin, and nails but didn't observe any dryness or thinning, and her thyroid was a normal size. There didn't seem to be any physical signs of a metabolic problem.

After checking everything off the list, Felitti sat down with her for a talk.

"Patty, what do *you* think is going on here?"

"You mean the weight?"

"Yes."

Her smile dimmed and she looked down at her hands.

"I think I'm sleep-eating," she said sheepishly.

"What do you mean?" Felitti asked.

"When I was a kid, I used to be a sleepwalker. I haven't done that for years, but I live alone and when I go to bed at night everything is clean and put away in the kitchen. Now, when I wake up in the morning, the pots and dishes are dirty, the boxes and cans are open. Somebody has obviously been cooking and eating, but I can't remember any of it. Since I'm the only person there and I'm putting on weight, I guess it's the only explanation."

Felitti nodded. It seemed a little wacky, possibly even a sign of some sort of psychopathology. Ordinarily, he'd refer her to mental-health services and focus his attention on addressing her physical health, but something stopped him. His recent conversation with Donna made him realize there were things that might be affecting his patients' success that he wasn't getting at with his usual questioning. He decided to follow this thread even though it seemed outside of his area of expertise.

"Patty, that you're sleep-eating explains the weight gain, but why are you doing it *now?*"

"I don't know."

"But why didn't this happen three years ago, or three months ago?"

"I don't know."

Felitti tried again. His work in infectious disease and epidemiology wouldn't let him stop with the surface explanation. There was usually a trigger event. Cholera didn't affect so many people in the Soho neighborhood of London because of bad luck; there was something tying together all the people who got sick, and that something was a contaminated well.

Felitti doubted that Patty had started sleep-eating for no reason.

"Think hard, Patty. What's been happening in your life? Why would you start sleep-eating now?"

She was quiet for a moment.

"Well, I don't know if it's related, but there's this man at work," she said, looking down again.

Felitti waited, and eventually Patty went on to explain that in her job as a nurse at a convalescent home, she'd been in charge of a new patient who kept hitting on her. He was much older and married, and he had remarked on how good she looked now that she'd lost all that weight. He'd been propositioning her ever since. At first, Felitti was perplexed. It didn't totally line up that this rather mild harassment (it was the eighties, after all) was enough to set her off in such an extreme way, but as he probed further, things became a lot more clear. Patty had a lengthy history of incest at the hands of her grandfather, starting when she was ten years old. This was also when she had begun to struggle with her weight.

After Patty left that day, Dr. Felitti realized that he couldn't ignore the similarities between her and Donna. Maybe it had just been a coincidence, but what stuck with him was the timing. Both patients had begun to gain weight as children immediately subsequent to incidents of abuse. Fast-forward a few decades; Patty's sudden weight gain coincided with being hit on by her patient. Felitti wondered if she might be subconsciously protecting herself from what must have seemed like a recurring trauma by gaining weight. What if he had been looking at this all wrong? He, as a doctor, had perceived a patient's weight to be the problem. What if it was actually a *solution?* What if his patient's weight was a psychological and emotional barrier, something protecting her from harm? That would go a long way toward explaining why his most successful patients, the ones who had peeled off that protective layer, were so desperate to put it back on.

Felitti suspected that he might have glimpsed a hidden relationship between histories of abuse and obesity. To get a clearer picture of that potential relationship, when he conducted his normal checkups and patient interviews for the obesity program, he now began asking people if they had a history of childhood sexual abuse. To his shock, it seemed as if every other patient acknowledged such a history. At first he thought there was no way this could be true. Wouldn't he have learned about this correlation in medical school? However, after 186 patients, he was becoming convinced. But in order to make sure there wasn't something idiosyncratic about his group of patients or about the way he asked the questions, he enlisted five colleagues to screen their next hundred weight patients for a history of abuse. When they turned up the same results, Felitti knew they had uncovered something big.

· · ·

Dr. Felitti's initial insight about the link between childhood adversity and health outcomes led to the landmark ACE Study. This was a prime example of doctors thinking like detectives, following a hunch and then putting it through its scientific paces. Beginning with just two patients, this research would eventually become both the foundation and the inspiration for ongoing work giving medical professionals critical insight into the lives of so many others.

After the initial detective work within his own department, Felitti started trying to spread the word. In 1990 he presented his findings at a national obesity meeting in Atlanta and was roundly criticized by his peers. One physician in the audience insisted that patients' stories of abuse were fabrications meant to provide cover for their failed lives. Felitti reported that the man got a round of applause.

There was at least one person at the conference who didn't think Dr. Felitti had been hoodwinked by his patients. An epidemiologist from the Centers for Disease Control and Prevention (CDC), David Williamson was seated next to Felitti at a dinner for the speakers later that night. The senior scientist told Felitti that if what he was claiming — that there was a connection between childhood abuse and obesity — was true, it could be enormously important. But he pointed out that no one was going to believe evidence based on a mere 286 cases. What Felitti needed was a large-scale, epidemiologically sound study with thousands of people who came from a wide cross-section of the population, not just a subgroup in an obesity program.

In the weeks following their meeting, Williamson introduced Felitti to a physician epidemiologist at the CDC, Robert Anda. Anda had spent years at the CDC researching the link between behavioral health and cardiovascular disease. For the next two years Anda and Felitti would review the existing literature on the connection between abuse and obesity and figure out the best way to create a meaningful study. Their aim was to identify two things: (1) the relationship between exposure to abuse and/or household dysfunction in childhood and adult health-risk behavior (alcoholism, smoking, severe obesity), and (2) the relationship between exposure to abuse and/or household dysfunction in childhood and *disease*. To do that, they needed comprehensive medical evaluations and health data from a large number of adults.

Fortunately, part of the data they needed was already being collected every day at Kaiser Permanente in San Diego, where over 45,000 adults a year were getting comprehensive medical evaluations in the health appraisal center. The medical evaluations amassed by Kaiser would be a treasure trove of important data for Felitti and Anda because they contained demographic information, previous diagnoses, family history, and current conditions or diseases each patient was dealing with. After nine months of battling and finally gaining approval from the oversight committees for their ACE Study protocol, Felitti and Anda were ready to go. Between 1995 and 1997, they asked 26,000 Kaiser members if they would help improve understanding of how childhood experiences affected health, and 17,421 of those Kaiser health-plan members agreed to participate. A week after the first two visits for this process, Felitti and Anda sent each patient a questionnaire asking about childhood abuse and exposure to household dysfunction as well as about current health-risk factors, like smoking, drug abuse, and exposure to sexually transmitted diseases.

The questionnaire collected crucial information about what Felitti and Anda termed "adverse childhood experiences," or ACEs. Based on the prevalence of adversities they had seen in the obesity program, Felitti and Anda sorted their definitions of abuse, neglect, and household dysfunction into ten specific categories of ACEs. Their goal was to determine each patient's level of exposure by asking if he or she had experienced any of the ten categories before the age of eighteen.

1. Emotional abuse (recurrent)

2. Physical abuse (recurrent)

3. Sexual abuse (contact)

4. Physical neglect

5. Emotional neglect

6. Substance abuse in the household (e.g., living with an alcoholic or a person with a substance-abuse problem)

7. Mental illness in the household (e.g., living with someone who suffered from depression or mental illness or who had attempted suicide)

8. Mother treated violently

9. Divorce or parental separation

10. Criminal behavior in household (e.g., a household member going to prison)

Each category of abuse, neglect, or dysfunction experienced counted as one point. Because there were ten categories, the highest possible ACE score was ten.

Using the data from the medical evaluations and the questionnaires, Felitti and Anda correlated the ACE scores with health-risk behaviors and health outcomes.

First, they discovered that ACEs were astonishingly common — *67 percent* of the population had at least one category of ACE and 12.6 percent had *four or more* categories of ACEs.

Second, they found a dose-response relationship between ACEs and poor health outcomes, meaning that the higher a person's ACE score, the greater the risk to his or her health. For instance, a person with four or more ACEs was *twice* as likely to develop heart disease and cancer and *three and a half* times as likely to develop chronic obstructive pulmonary disease (COPD) as a person with zero ACEs.

. . .

Given what I'd seen in my patients and in the community, I knew in my bones that this study was dead-on. It was powerful evidence of the connection that I had seen clinically but had never seen substantiated in the literature. After reading the ACE Study, I was able to answer the question of whether there was a medical connection between the stress of childhood abuse and neglect and the bodily changes and damage that could last a lifetime. It seemed clear now that there was a dangerous exposure in the well at Bayview Hunters Point. It wasn't lead. It wasn't toxic waste. It wasn't even poverty, per se. It was childhood adversity. And it was making people sick.

. . .

One of the most revealing parts of the ACE Study was not *what* it investigated but *who* it investigated. Many people might look at Bayview Hunters Point and see the rates of poverty and violence and the lack of health care and say, "Of course those people are sicker; that makes sense." After all, that's what I learned in public-health school. Poverty and lack of adequate health care are what really drives poor health outcomes, right?

This is where the ACE Study comes in and shakes things up, showing us that the dominant view is missing something big. Because *where* was the ACE Study conducted?

Bayview? Harlem? South-Central Los Angeles?

Nope.

Solidly middle-class San Diego.

The original ACE Study was done in a population that was 70 percent Caucasian and 70 percent college-educated. The study's participants, as patients of Kaiser, also had great health care. Over and over again, further studies about ACEs have validated the original findings. The body of research sparked by the ACE Study makes it clear that adverse childhood experiences in and of themselves are a risk factor for many of the most common and serious diseases in the United States (and worldwide), regardless of income or race or access to care.

• • •

The ACE Study is powerful for a lot of reasons, but a big one is that its focus goes beyond behavioral or mental-health outcomes. The research wasn't conducted by a psychologist; it was conducted by two internal medicine doctors. Most people intuitively understand that there's a connection between trauma in childhood and risky behavior, like drinking too much, eating poorly, and smoking, in adulthood (more on that later). But what most people *don't* recognize is that there is a connection between early life adversity and well-known killers like heart disease and cancer. Every day in the clinic I saw the way my patients' exposure to ACEs was taking a toll on their bodies. They may have been too young for heart disease, but I could certainly see the early signs in their high rates of obesity and asthma.

• • •

Along with my excitement at finding the ACE Study's demonstration of links between adversity and disease came a wave of indignation: *Why was I only hearing about this now?* This study was clearly a game-changer, yet I hadn't learned about it in med school, public-health school, or even residency. Felitti and Anda published their initial ACE findings in 1998, and I didn't read them until 2008. Ten years! And still this important science hadn't been translated into clinical tools I could use to improve my patients' health. How could that be possible?

When I talked to Felitti years later, he mentioned attacks on parts of the paper by various colleagues. While Felitti and Anda successfully refuted the criticisms, the work never seemed to gain traction. In fact, it almost seemed to disappear, which is kind of crazy when you think about what the study revealed. Dr. Anda's colleagues at the Centers for Disease Control were agog, telling him that the magnitude of the increased likelihood of disease was the sort seen only a couple of times in a researcher's career. A critical piece of their findings was the dose-response relationship; for example, the more cigarettes you smoke and the more years you smoke them, the higher your odds of developing lung cancer. The ACE Study strongly establishes a dose-response relationship, which is an important step toward demonstrating causality. A person with an ACE score of seven or more has *triple the lifetime odds* of getting lung cancer and *three and a half times the odds* of having ischemic heart disease, the number one killer in the United States. If a large study like Felitti and Anda's came out tomorrow saying that exposure to cottage cheese tripled your lifetime chances of cancer, the Internet would break and the dairy lobby would hire a crisis-management firm.

. . .

So what gives? Why hadn't I heard of this study before? Why wasn't I listening to stories about it on NPR and watching Dr. Felitti be interviewed by Oprah? I can see now that there were at least three reasons.

The first has to do with a misconception concerning the ACE Study itself, the belief of some that the increased risks had everything to do with behavior. As I said earlier, many people assume they understand the adversity-health connection. The popular thinking goes that if you live in poverty or have a rough childhood, you inevitably cope by drinking and smoking and doing other risky things that damage your health. But if you're smart and strong, you rise above what you were born and raised with and leave the bad things behind. At first, this construct seems to make sense, but remember, at one point it made perfect sense that the Earth was flat.

Fortunately, some smart scientists decided to test the behavioral assumption. They looked at the association between ACEs and heart and liver disease and did some very complicated analyses to assess how much of the disease was due to the effects of health-damaging behaviors like smoking, drinking, physical inactivity, and obesity. It turned out that "bad behavior" accounted for only about 50 percent of increased likelihood for disease. In a way that's good news, because it means that if a person is exposed to ACEs and he is careful to avoid smoking, physical inactivity, and other health-damaging behaviors, he can protect himself from about 50 percent of the health risk. But it also means that even if he doesn't engage in any health-damaging behaviors, he's *still* more likely to develop heart or liver disease.

Felitti's patient Patty is a good example. She was severely obese and a self-described sleep-eater, so obviously her behaviors caused her obesity, which caused her later health problems, right? Not so fast. After dropping out of Dr. Felitti's program initially, she returned later, asking for more help with her weight problem. Over the years she would lose the weight and put it back on over and over again, even after bariatric surgery. Sadly, Patty died at the age of forty-two of pulmonary fibrosis, an autoimmune condition that damages the lung tissue and makes breathing difficult and, eventually, impossible. But obesity is not the cause of pulmonary fibrosis. Patty didn't smoke and she had never been exposed to any known pulmonary toxins like asbestos. Having an ACE score of two or more doubles someone's likelihood of developing an autoimmune disease. Patty's ACEs were likely her biggest risk factor, yet neither she nor her doctors knew it.

In the United States, the culture puts a lot of stock in personal responsibility. The lifestyle choices you make do have a huge impact on your health; so-called bad behavior *does* result in increased risks to your health, and there's no disputing that. But the ACE Study shows us, yet again, that it's not the whole story.

The second reason I hadn't heard of Felitti and Anda's work in medical school, and maybe the most potent, is that this is scary, emotional stuff. It's one thing to take a cold, calculating look at your cottage-cheese consumption over the past decade, but it's another to revisit trauma and abuse. I bet everyone reading this book can think of someone who grew up with a family member who suffered from mental illness or who had a parent who drank too much, or who was emotionally abusive, or who believed that sparing the rod spoiled the child. In any group you might find yourself in — a classroom, a professional conference, a wedding party, the U.S. Congress — if everyone's ACE score was suddenly revealed, it would show pretty clearly that this is an issue that touches many of us. But most of us don't like to think about the sad, upsetting things that have happened in the past. It's possible that we marginalize the impact of trauma on health because it *does* apply to us. It's hard, after all, to accept that there might be biological implications that persist whether people are sinners or saints. Maybe it's just easier to see it in other zip codes.

The last reason why the ACE Study didn't catch fire in the medical and scientific communities in 1998 can be best explained as scientific gaps. The study showed that adversity was bad for your health, but although Felitti and Anda had exposed the *what,* they were unable at the time to answer the *how.*

Lucky for me, there had been ten years of intervening research that had slowly but surely filled in those scientific gaps.

Now what I needed to do was return to the Hayes lab and Sarah P. and dig deeper into that *how.* In my gut, I had a strong sense of what puzzle pieces fit in the ACE Study's scientific gaps. Identifying and demonstrating that the stress-response system was the biological mechanism behind adversity's role in health was going to be the fun part. I'd have to jump back into those journals and hit up some medical conferences, but now I had the ACE Study to guide me. I could use its language in my searches, interrogate its authors for clues, and even start collecting my own ACE data at the clinic. The realization that this was bigger than my patients, bigger than Bayview, made my heart pound. Adversity's detrimental impact on health had all the hallmarks of a public-health crisis hidden in plain sight.

Before I met Diego or even knew about ACEs, I had hope for Bayview. I knew problems there were amplified but that their solutions would be too. On our first day at the clinic, I told my staff that if we could successfully treat people here, we could treat people anywhere.

II

Diagnosis

4

The Drive-By and the Bear

IT WAS CHILLY, WHICH was typical for a December night in San Francisco, but as I walked down Mission Street with my friends, I remember hugging myself to try and warm up. Home for the holidays from public-health school in Boston, I'd optimistically left the house without a coat. I told myself to be grateful; after all, it wasn't snowing. I was so amped up from a night out with old friends that my poor clothing choice was just another thing to laugh about. The four of us all talked simultaneously, our voices drowning out the sounds of the city as we made our way back to the car. We lingered on the corner of Nineteenth Street and Mission, not wanting to part ways as the evening came to an end. None of us noticed the red car slowing down across the street from us until seconds later when we heard *Pop! Pop! Pop!* As the car peeled out toward Twentieth Street, my friend Michael initially laughed it off.

"It's just some stupid kids playing with firecrackers," he said, collecting himself after the startle. But a few moments later, he picked up on an uneasy vibe and ushered us toward the car. "We gotta get out of here. Something's going down."

We were almost to Michael's vehicle when we saw the man lying on the sidewalk. Three men I assumed were his friends were a few feet away, yelling and punching in the windows of the cars parked on that side of the street.

"Oh my God," cried my cousin Jackii, "he's been shot!"

As if by reflex, I headed toward the victim, not noticing that my friends were running in the opposite direction.

"Nadine!" Michael said. He grabbed at my arm, but he was too late. I dropped to my knees when I reached the man's side. All I could think about was saving his life. I had finished medical school the year before; my doctor instincts now took over. As I got a good look at his face, I could see that despite his size, he was still a boy. He couldn't have been more than seventeen. There was an entry wound above his right eyebrow, and because he was lying on his side I could see a fist-size exit wound at the back of his head. My inner narrator began calling out a status report like we had been trained to do in the trauma bay: "Gunshot wound to the head! No other signs of penetrating trauma!"

In the movies, the guy would have been out cold, but in real life he was throwing up on himself. I'd seen a lot of scary stuff in hospitals, but this was different. Time seemed to slow down and I found myself going on autopilot. I kept checking things off the list that I had learned in medical school: *ABCD — airway, breathing, circulation, disability. Keep his airway clear. Make sure he's breathing. Check his pulse. Maintain the position of his cervical spine in case his neck is broken.* At the same time, a voice in the back of my head kept telling me that I wasn't in the safety of the emergency department — there was no security guard at the door, and the red car could come back! My heart was pounding and my hands shook. Every cell in my body was telling me to get the hell out of there, but I stayed by his side until the paramedics arrived.

Hours later, as we sat in the Mission district police station giving our best descriptions of what we had seen, we got the news that he didn't make it. It was a heartbreaking end to the evening, but I knew there wasn't anything else I could have done. After I got home that night, I couldn't sleep. In the following weeks and months, every time I saw a red car approaching quickly or heard a car backfire, I felt transported back to the fear I felt that night. Physically, I would have the same responses: my heart would pick up its pace, my eyes would dart around, and I'd feel tightness in my stomach. I see now that my biology was reacting to an unusually high level of stress by temporarily linking red cars with danger. My body was remembering what happened and sending a flood of stress hormones shooting through my system in case the red car *now* was as dangerous as the red car *before*.

My body was doing what it was designed to — keep me out of harm's way.

Day to day, your brain has to process *a lot* of information — trees creaking in the wind overhead, dogs barking next door, the wall of air hitting your face as the subway car hurtles by — and interpret the risk. In order for humans to survive, the brain and body had to come up with efficient ways of processing information, and the stress-response system is one of them. If a little kid touches a hot stove, his body remembers. Biochemically, it tags or bookmarks the stove (and all the stimuli associated with it) as being dangerous, so the next time the boy sees someone turning on the burners, his body sends him all kinds of warning signs: vivid memories, muscle tension, and rapid pulse. Usually, this is enough to dissuade him from doing the same thing again. In this way, our bodies are trying to protect us, which makes a lot of sense. The prehistoric creatures that didn't evolve that mechanism didn't live to reproduce.

But the stress response can do its job a little *too* well sometimes. This happens when the response to stimuli goes from adaptive and lifesaving to maladaptive and health-damaging. For example, almost everyone knows that soldiers sometimes come back from the front lines with posttraumatic stress disorder. This condition is an extreme example of the body remembering too much. With PTSD, the stress response repeatedly confuses current stimuli with the past in such a dramatic way that it becomes hard for these vets to live in the present. Whether it's a B-52 bomber in the sky or a commercial airliner overhead taking tourists to Hawaii, their bodies feel the same — *in mortal danger*. The problem with PTSD is that it becomes entrenched; the stress response is caught in the past, stuck on repeat.

For me, the specific trigger of a red car would eventually decouple from my body's most ancient defense mechanism and stop being interpreted by my brain as a threat. Now when I see one go by me on a city street I don't flinch. What I didn't know until years later was *why*. Why was my body able to recover from that intense instance of stress? What made the sensory connection between the red car and the biological reaction of my stress response fade? I wouldn't think to ask those questions until many years later, when Diego set me on my path.

• • •

In the months after first discovering the ACE Study, I once again dived into the research. I found that some robust and incredibly exciting advances had been made in the biology of stress and its impact on child health and development. What I know now is that what happened in my body that night in the Mission district is the same thing that happens to my patients' bodies when they experience a whole host of adversities ranging from abuse to abandonment. The body senses danger, and it sets off a firestorm of chemical reactions aimed to protect itself. But most important, *the body remembers.* The stress-response system is a miraculous result of evolution that enabled our species to survive and thrive into the present. We all have a stress-response system, and it is carefully calibrated and highly individualized by both genetics and early experiences. What makes the stress response of a child with zero ACEs different from Diego's stress response is a complicated question that we will begin to unravel, but it all starts with the same system. When it's in good working order, it can help save your life, but when it's out of balance, it can shorten it.

Stress Response

Flipping through a magazine in the checkout line or circling the Internet vortex, you've probably come upon stories of superhuman strength: the father who lifts the car that was pinning his child (maybe an urban legend?) or the woman who fought off the mountain lion that was mauling her husband (that one really is true). Even more cinematic, there's the average Joe turned hero who crosses the battlefield to save his buddy despite having two bullets in him. If you've ever wondered what makes a person able to achieve such feats, I can tell you it's not the daily bowl of Wheaties — it's the elegantly designed, evolutionarily imperative stress-response system. Essentially, it works like this: Imagine you're walking in the forest and you see a bear. Immediately, your brain sends a bunch of signals to your adrenal glands (perched on your kidneys) saying, *"Release stress hormones! Adrenaline! Cortisol!"* So your heart starts to pound, your pupils dilate, your airways open up, and you are ready to either fight the bear or run from the bear. That's the response commonly known as *fight or flight*. It has evolved over millennia to save your life. Another lesser-known way your body might respond is by freezing, in the hopes that the bear will think you're a rock. For that reason, some people use the term *fight, flight, or freeze*, but to keep it simple, I'm just going to use fight or flight.

• • •

To understand and appreciate how the stress-response system can go wrong or, as we doctors say, "become dysregulated," there are a few basic things to know about what happens when it goes right. Be warned that this biological system is one of our species' oldest and most complicated. People take entire courses on this stuff and still walk away slightly confused. I'll try to keep it as simple and accurate as possible.

Here are the main players involved.

- The amygdala: the brain's fear center
- Prefrontal cortex: the front part of the brain that regulates cognitive and executive function, including judgment and mood and emotions
- Hypothalamic-pituitary-adrenal (HPA) axis: initiates the production of cortisol (longer-acting stress hormone) by the adrenal glands
- Sympatho-adrenomedullary (SAM) axis: initiates the production of adrenaline and noradrenaline (short-acting stress hormones) by the adrenal glands and brain
- Hippocampus: processes emotional information, critical for consolidating memories
- Noradrenergic nucleus in the locus coeruleus: the within-the-brain stress-response system that regulates mood, irritability, locomotion, arousal, attention, and the startle response

Now let's go back to the forest.

When you see the bear, your amygdala immediately sounds the

alarm telling your brain to be afraid because bears are *scary!* Your brain then activates your SAM and HPA axes, triggering the fight-or-flight response. Signals from the SAM axis travel along nerves from the brain to the adrenal glands telling them to make adrenaline, which is responsible for many of the feelings that we associate with being terrified. Adrenaline causes the heart to beat stronger and faster, sending blood to all the places that need it. It causes your airways to open so that you can take in more oxygen. It raises your blood pressure and shunts blood toward your skeletal muscles (necessary for running and jumping) and away from that tiny muscle that holds your bladder closed, which is why scared people feel like they're going to pee in their pants and sometimes do. It also converts fat to sugar for energy.

The SAM axis also activates the noradrenergic nucleus of the locus coeruleus, which, as I like to say, is the scientific term for the part of the brain responsible for "I don't know karate but I do know c-razy!" This is the within-the-brain stress-response center, and it gets you amped up. (Picture Oakland Raiders fans after a winning game or, worse, a losing game.) Adrenaline and noradrenaline are powerful stimulants, designed to help you think more clearly so that you can figure out the quickest path to safety. They also create feelings of euphoria, that adrenaline rush that makes you think you can conquer the world. But, like everything body-chemistry-related, it's all about balance. A graph of the response of the prefrontal cortex (the part of the brain responsible for reason, cognition, and judgment) to adrenaline and noradrenaline looks like an inverted U — a little bit improves functioning, but too much will mess up your ability to focus.

Now your heart is pumping, your muscles are primed, and you are feeling ready to fight. If you were to stop and think about it, fighting a bear might seem like a bad idea. After all, grizzly bears can weigh up to seventeen hundred pounds. They have huge teeth and fearsome claws. Odds are that you aren't going to fare very well. That's why when you're *really* scared, your fear center temporarily shuts down the thinking part of your brain — because you need to defy those odds. Your life depends on it. So the amygdala activates neurons that link to the prefrontal cortex and temporarily turns it off, or at least turns it way, way down. The SAM axis is a very short-acting (seconds or minutes) response that primes your body by making available what you need most: blood, oxygen, energy, and chutzpah.

At the same time, the HPA axis triggers hormones in the brain that let loose a cascade of chemical messengers that ultimately result in the release of some longer-term stress hormones, most notably cortisol. Imagine if you lived in a forest where there were lots of bears. After the first one or two encounters, your body would want to become more efficient at responding to the bear problem. Essentially, cortisol helps the body adapt to repeated or long-term stressors, like living in bear-infested woods or handling prolonged food shortages. Some of the effects of cortisol are similar to those of adrenaline — it raises blood pressure and blood sugar, inhibits cognition (clear thinking), and destabilizes mood. It also disrupts sleep, which makes a lot of sense if you are living in a forest full of bears — better to be a light sleeper. Unlike adrenaline, which can decrease appetite and stimulate fat burning, cortisol stimulates fat accumulation and also triggers the body to crave high-sugar, high-fat foods. Think about your last breakup. If you're wondering why you couldn't sleep and were tunneling your way to the bottom of a pint of Häagen-Daz, that's cortisol. High levels of cortisol can inhibit reproductive function because if you are living in the forest next to bears, isn't it better to wait to have kids until you move to a safer part of the woods?

One not-so-obvious but incredibly important function of the stress response is activating the immune system. After all, if you are fighting a bear, he might get in a couple of licks. If he does, you want your immune system primed to heal, meaning that it's ready to bring inflammation to the area in order to stabilize the wound and keep you fighting long enough to beat the bear or get away.

Once you do get away and are back in the safety of your cave, both the SAM and the HPA axes are designed to shut themselves down. The body uses a sort of stress thermostat called feedback inhibition that triggers the stress response to turn itself off once it has done its job. High levels of adrenaline and cortisol feed back to the parts of the brain that initiate the stress response and turn them off. What an incredibly evolved system! Especially if you live in a forest and there are bears. But what happens when you can't experience safety in your cave because the bear is living in the cave with you?

Living with the Bear
(aka Dysregulated Stress Response)

Over and over again in my practice I saw kids who had experienced terrifying situations. For one patient, the bear was his dad who verbally demeaned and physically abused his mom. For another, it was his mom when she didn't take her psychiatric medications and left the kids uncared for, often in dangerous situations. I'll never forget the fourteen-year-old girl for whom the bear was the very neighborhood she lived in after she was hit by a stray bullet walking home from school.

For many of my patients, the stress response was activated dozens and sometimes hundreds of times a day. I knew that if I wanted to get to the source of the problem for Diego and other patients, I needed to understand exactly when and how the stress response begins working against the body. What happens to children's brains and bodies when they are exposed to such high doses of adversity? Fortunately, some smart scientists were asking the same question.

During one of my trips down the research rabbit hole, I found some great work by Jacqueline Bruce, Phil Fisher, and colleagues. In a 2009 study, they set out to determine if the adverse experiences of preschool-age foster children had an effect on the functioning of the stress-response system, specifically the HPA axis. To do this, they analyzed the cortisol levels of 117 foster kids and 60 low-income kids who were *not* maltreated. What they found reinforced what I suspected about my own patients: the foster kids showed dysregulated cortisol levels in comparison to the kids who had not experienced the same adverse experiences.

It turns out that cortisol has a predictable daily pattern: it's high in the morning to help wake you up and get you ready for the day and then gradually decreases, reaching its lowest point in the evening, just when you need to go to sleep. As a result, it's possible to determine if someone's cortisol pattern is disrupted. Fisher and Bruce found that children who had experienced maltreatment had higher overall cortisol levels as well as a disruption of the normal daily pattern of cortisol secretion. The morning peak wasn't quite as high and the daily decline was not as steep, leading to higher levels in the evening and higher total daily cortisol.

One interesting part of the foster kids study was that the control group did not consist of children who were demographically all that different from the experimental group in terms of parental education and income. The major differences were that the control-group kids were all living with at least one parent, had not had any previous contact with child services, and were not maltreated. Undoubtedly, the low-income kids in the control group *had* been exposed to at least some level of adversity in their lives, and yet their cortisol levels were not abnormal. This sheds some light on how some children can experience stress without that stress tipping over into dysregulation.

We all know that adversity, tragedy, and hardship are a part of life. As much as we'd like to shield our children from illness, divorce, and trauma, sometimes these things happen. What the research tells us is that those daily challenges can be overcome with the right support from a loving caregiver.

Fisher went on to work with the National Scientific Council on the Developing Child as part of an ambitious effort to pull together the science of how early adversity affects the developing brains and bodies of children. The council, too, found that a dysregulated stress-response system was at the core of the problem.

The main issue is that when the stress response is activated too frequently or if the stressor is too intense, the body can lose the ability to shut down the HPA and SAM axes. The term for this is *disruption of feedback inhibition*, which is a science-y way of saying that the body's stress thermostat is broken. Instead of shutting off the supply of "heat" when a certain point is reached, it just keeps on blasting cortisol through your system. This is exactly what Fisher and Bruce were seeing in the foster kids.

Ultimately, the council described three different kinds of stress responses:

- Positive stress response *is a normal and essential part of healthy development, characterized by brief increases in heart rate and mild elevations in hormone levels. Some situations that might trigger a positive stress response are the first day with a new caregiver or receiving an injected immunization.*

A good example of positive stress is one that many athletes can relate to: pregame jitters. The moment before a big race, a track star might feel a rush of nervousness. Physically, her heart rate is amped up and she's got butterflies in her stomach. But the increase in adrenaline is doing important work. The track star is taking in more oxygen, shunting more blood to her muscles, and heightening her focus. When the starter's gun goes off, she's ready for it.

- Tolerable stress response *activates the body's alert systems to a greater degree as a result of more severe, longer-lasting difficulties, such as the loss of a loved one, a natural disaster, or a frightening injury. If the activation is time-limited and buffered by relationships with adults who help the child adapt, the brain and other organs recover from what might otherwise be damaging effects.*

Lots of kids wet the bed when they are little but grow out of it. An example of a tolerable stress response would be a child who reverts back to bedwetting after his parents' divorce. The split isn't acrimonious, and while the dad moved out, both adults are committed to co-parenting and understand that their child needs stability and extra support. As a result of that buffering of the child's stress, he stops wetting the bed after a few months. Like my drive-by-induced stress, the effects are temporary if a solid support network is in place.

- Toxic stress response *can occur when a child experiences strong, frequent, and/or prolonged adversity — such as physical or emotional abuse, neglect, caregiver substance abuse or mental illness, exposure to violence, and/or the accumulated burdens of family economic hardship — without adequate adult support. This kind of prolonged activation of the stress-response systems can disrupt the development of brain architecture and other organ systems, and increase the risk for stress-related disease and cognitive impairment, well into the adult years.*

There was no question in my mind that Diego was experiencing a toxic stress response. Beyond Diego's sexual abuse when he was four, he and his family had dealt with other hardships that also put his system under strain. Diego's dad clearly had a drinking problem, and his mother was suffering from depression. Neither was able to be an adequate stress buffer for him. Diego's constellation of symptoms were directly in line with what we know happens when there is a prolonged activation of the stress-response system without adequate support.

. . .

Healthy development of the stress-response system requires that a child experience both positive and tolerable stress. This allows the SAM and HPA axes to be calibrated to react normally in the face of stressors. But for every ACE a child has, the risk of tolerable stress tipping over into toxic stress increases, as the system responds more frequently and intensely to multiple stressors.

Just like tadpoles, children are particularly sensitive to repeated stress activation. High doses of adversity affect not only the brain structure and function but also the developing immune system and hormonal systems, and even the way DNA is read and transcribed. Once the stress-response system gets wired into a dysregulated pattern, the biological effects ripple out, causing problems within individual organ systems. Because the body is like one big, intricate Swiss watch, what happens in your immune system is deeply connected to what happens in your cardiovascular system. Next we'll see the downstream effects of a stress-response system that has gone off the rails.

5

Dynamic Disruption

IF YOU WANT TO understand how a child's stress response is working, try walking into the examination room with a tray full of needles and telling him it's time for shots. By now, it seemed like I could almost guess the ACE score of a patient by the amount of commotion that took place when my nurse went in to give the vaccinations. We'd seen it all: screaming, kicking, biting, kids literally trying to climb the walls to get away from the needles. One patient got so upset he vomited on my white coat. Another ran out of the exam room and made it all the way down the block before we caught her. These extreme displays of fear were not your ordinary needle-phobic reactions; they were full-blown bear-in-the-woods reactions. Coincidentally, this natural stress-response provocation challenge gave us an opportunity to test the second, equally important ingredient for toxic stress — the caregiver's ability to act as a buffer. The kids who had the worst responses were also the ones whose caregivers were the *least* likely to hug, kiss, sing to, or otherwise soothe their child. We heard a lot of "Hold him down!" and "I don't have time for this, I have to be back at work in a half hour."

Observing that phenomenon and suspecting a correlation was one thing, but I needed to find a way to rigorously evaluate not just *whether* ACEs had an impact on my patients, but *how*. Dr. Victor Carrion, a child psychiatrist and the director of the Early Life Stress and Pediatric Anxiety Program at Stanford University Medical Center, soon became an ally.

There is still a lot we don't know about how stress affects the brain, but every day, promising studies show us more and more. We know as much as we do about toxic stress's impact on the brain because of important research like Dr. Carrion's at Stanford.

Carrion had been working for a long time with kids who were exposed to high doses of adversity. Previous research in adults showed that high levels of cortisol were toxic to the hippocampus, but Dr. Carrion decided to look specifically at kids. Thanks to MRI technology, he was able to peek inside their brains and see cortisol's impact on kids who had experienced trauma. What's so compelling to physicians about Dr. Carrion's work is that it told the story in a language that we doctors were accustomed to hearing. When you put a kid who had experienced adversity in an MRI machine, you could see *measurable changes* to the brain structures.

For the study, Carrion and his team recruited patients from various local health services. The criteria were that they had to have been exposed to trauma, were between ten and sixteen years old, and had PTSD symptoms. Most of the kids had experienced multiple traumatic events — witnessing violence or suffering physical abuse or emotional abuse. Many of them were living in poverty. The control group had no history of trauma but were comparable to the experimental group in terms of income, age, and race. In preliminary interviews, the researchers asked the kids or their caregivers about PTSD symptoms and hyperarousal symptoms like difficulty sleeping, irritability, and trouble concentrating, to name just a few. Then they did an MRI and checked each kid's salivary cortisol four times a day. Once the brain scans were in, they looked at the size of each child's hippocampus by measuring the volume in 3-D. They found that the more symptoms a kid had, the higher his cortisol levels were and the smaller the volume of his hippocampus. After the first measurement of the hippocampus, they measured the same kids again twelve to eighteen months later and found their hippocampi were even smaller. Despite the fact that these kids were no longer experiencing trauma, the parts of their brains responsible for learning and memory were still shrinking, showing us that the effects of earlier stress were still acting on the neurological system.

Dr. Carrion agreed with me that it was important to assess my entire population of patients for the effects of toxic stress, and he was as interested as I was in the results. We decided our focus would be on the association between ACE scores and two of the most common issues I saw in my patients: obesity and learning/behavior problems. After a careful review of each patient's chart, my research assistant Julia Hellman assigned everyone an ACE score. We even had another reviewer from Stanford review and score a random sampling of our patient charts to make sure that our scoring was accurate.

At first, the ACE scores of our study population of 702 patients looked a lot like Felitti and Anda's: 67 percent of our kids had experienced at least one ACE, and 12 percent had experienced four or more. I have to admit that I was surprised that our numbers weren't higher. After all, Bayview was a pretty rough neighborhood. I knew the questions that Felitti and Anda asked didn't cover everything my patients had been through, like community violence or having a family member deported, both common occurrences in the lives of my kids. But still, I expected our patients in Bayview to have experienced more ACEs than the Kaiser population. But then I had a forehead-slapping realization. Felitti and Anda had done their study among *adults*. The mean age of their patients was fifty-five. The subjects were asked to recall the number of ACEs experienced by the time they were eighteen. In our study, the mean age was *eight*. Many of our kids would likely have more ACEs before they reached their eighteenth birthdays. We also had to consider that it was the caregivers, not the children themselves, who reported the adverse experiences we were charting, and these caregivers might not have reported adversity accurately because of shame or fear or because "we just don't talk about those things."

Apart from these revelations, the profound discovery was that our patients with four or more ACEs were *twice* as likely to be overweight or obese and *32.6 times* as likely to have been diagnosed with learning and behavioral problems. When our statistician from Stanford first called to tell me how these numbers shook out, I was overwhelmed by a mix of emotions — elation at making an important discovery and a profound aching in my heart for all the kids who were struggling in school but being told that they had ADHD or a "behavior problem" when these problems were directly correlated with toxic doses of adversity.

The reason this is so important is that an accurate diagnosis *should* tell physicians the underlying biological problem so they can provide the best treatment and the most likely prognosis. For example, if a patient is found to have cancer in his liver, it's critical for his doctors to know whether the cancer originated in the liver or metastasized from the prostate or somewhere else in the body; the treatments and prognoses for various cancers are different, even though the initial physical finding may be the same. Currently, ADHD is a diagnosis based entirely on symptoms. If you remember, the criteria include inattention, impulsiveness, and hyperactivity, but the *Diagnostic and Statistical Manual of Mental Disorders* doesn't say a word about the underlying biology. What it does say is that if these same symptoms are associated with a *different* mental disorder, like schizophrenia, then it's no longer ADHD. Similarly, if we see impulsivity and hyperactivity but discover that those symptoms are caused by a brain tumor, we can't diagnose ADHD.

From Felitti and Anda's research, I was beginning to understand that the prognosis of toxic stress, the long-term risks that my patients faced, looked very different from run-of-the-mill ADHD. We have a ways to go before we fully understand whether the behavioral symptoms of toxic stress represent a totally different diagnosis. Part of the problem has been that, unlike ADHD, the diagnosis of toxic stress doesn't yet exist in the medical literature.

This clinical pattern has an echo in recent medical history. In the 1980s, the medical world was confronted with a new epidemic. People would go to see their doctors, complaining of rashes and sores. They would make their way to emergency rooms with tuberculosis and hepatitis C. Even more baffling, they showed up in droves with Kaposi's sarcoma, a rare type of cancer that attacks the skin, mouth, and lymph nodes. For a while, no one suspected there was any connection among these health problems because they were known quantities. Doctors did what they were trained to do and treated the sores, the hepatitis, the cancer. But symptomatic patients kept coming in at higher rates than anyone had ever seen. So doctors felt they had to get better and better at treating things like sores, hepatitis, and Kaposi's sarcoma — a strategy that didn't touch the underlying problem. These patients kept getting sicker and sicker. Now we know that sores, tuberculosis, and Kaposi's sarcoma were all indicators of a more significant underlying problem, an infection compromising the entire immune system. These were AIDS-defining diseases; they were conditions that needed intervention and symptoms pointing to an underlying biological problem with a very different prognosis and treatment: HIV/AIDS.

So when I looked at my patients with high ACE scores, I couldn't help but think that if I treated *just* the asthma or the obesity or the behavior problem, I was a really poor student of history. We know from the research that the life expectancy of individuals with ACE scores of six or more is twenty years shorter than it is for people with no ACEs. For a patient with a high ACE score, it may not be the obesity that shortens his or her life but the underlying toxic stress that the obesity is signaling. To treat the root of the problem I had to look at both stories my patients' symptoms were telling me: the story on the surface and the story underneath. So when a patient named Trinity walked through the door with a chief complaint of ADHD, I was ready for her.

I was starting to get a reputation in the area for being the type of doctor who wouldn't just slap a prescription for Ritalin on the table. People brought their kids to me when they wanted someone to take a closer look. But before I knew just how close to look in Trinity's case, I had to know her ACE score. After the chart review of our initial 702 patients, I began asking about exposure to adversity for *all* of my patients to better understand their health risks. Just like height, weight, and blood pressure, the ACE score became another vital sign for my regular medical exams. With Trinity's complaint of learning and behavior problems, if her ACE score had been zero, a standard ADHD workup would have been warranted. But now I knew that if a patient had four or more ACEs, she was thirty-two times as likely to have learning or behavior problems, which suggested that the underlying issue was probably not ordinary ADHD. In those cases, I was convinced that the problem was chronic dysregulation of the stress-response system, which inhibited the prefrontal cortex, overstimulated the amygdala, and short-circuited the stress thermostat — in other words, toxic stress. When I flipped through Trinity's chart, I saw that she had an ACE score of six.

When I first walked into the exam room and met Trinity, I had an immediate childhood flashback. Before my family moved to the United States from Jamaica, I started first grade at Hope Valley Elementary School in Kingston. It was there that I found the thing that my household of four brothers lacked — other girls to play with. There was a gaggle of older girls who adopted me and taught me critical lessons like how to jump rope and climb the jungle gym with a skirt on. I would beg my mom to braid my hair into neat plaits just like theirs. They were long-limbed and lean, with cocoa-brown skin and bright white teeth. Trinity would have fit right in, down to the school uniform she wore — a crisp white cotton short-sleeved shirt and a navy-blue knee-length wool skirt. I noticed she was tall for her eleven years and slimmer than average, though I doubted she walked three miles to school every day like the girls from my childhood. Trinity was sitting quietly with her aunt, eyes scanning the room. She was polite and obedient and super-sweet. Before I even had to ask, Trinity's aunt launched into the story behind her niece's ACE score.

Trinity's mom was a heroin addict who made only unpredictable, cameo appearances in her daughter's life. She'd roll into town out of nowhere and pick Trinity up to go shopping. But what "shopping" really meant was hitting up department stores and using her daughter as a decoy while she boosted clothing and shoes. Trinity's aunt had stopped allowing the mom to visit when she found out Trinity herself had begun lifting lip gloss and other small items when she was out with her mom. Since then Trinity had been having major problems in school, and her teachers were at the end of their rope. Beyond the learning issues, she was having difficulty with emotional regulation. She'd act out and get into trouble with the kid next to her, and she couldn't sit still for more than five minutes. Sometimes she even ran out of the classroom.

As with most of my kids, I would never have suspected the trouble from the calm way Trinity behaved herself in my exam room. But I began my physical exam with my toxic stress lens on, giving Trinity an even more careful once-over than I would have for a kid with zero ACEs, kind of like how, if a patient lives with two parents who are heavy smokers, I sure as hell give an extra-close listen to that kid's lungs. Knowing Trinity was at higher risk for a whole host of things, I listened hard to her lungs (no wheezing). I looked at her skin (it was warm and soft with no dryness or flaking). I looked at her hair (there was breakage at the edges, but that was a common finding among African American girls, depending on their hairstyle). Nothing seemed terribly out of the ordinary — until I got to her heart.

Most people know that a regular heartbeat (no skipping or murmurs) is something that doctors look for, but what we're also looking for is how hard it beats. When I laid my stethoscope on Trinity's chest, I had to pause to readjust my earpieces. It was as though the volume on her heartbeat was turned up just a little higher than normal. It was subtle, but instead of the soft *lub-dub* I expected to hear, it was more like a LUB-DUB. I took off my stethoscope and looked at her for a moment. Then I gently laid my hand on her chest. No, I wasn't imagining it. Not only did her heartbeat sound louder than normal, it felt stronger than normal as well. The heartbeat question combined with her slimness was enough of a red flag for me to send her for an EKG.

The next day, the EKG confirmed the abnormality with her heart. According to the results, it was beating faster and the muscle was working harder than normal. The cardiologist who interpreted the EKG included a note that reinforced my suspicions: *possible Graves' disease.* Slim builds and strong heartbeats (as well as hair breakage) can be signs of Graves', which is an autoimmune disease that results in the thyroid gland being overstimulated. Unlike the example I gave earlier of hypothyroidism (when the thyroid gland doesn't make enough thyroid hormone), Graves' disease is a case of *hyper*thyroidism, where the thyroid gland makes too much thyroid hormone. If you remember, adults with hypothyroidism gain weight easily and can be somewhat lethargic. By contrast, people with Graves' disease are often hyperactive and can't seem to keep weight on.

In Europe, hyperthyroidism is often called Basedow's disease, after Karl Adolph van Basedow, the German physician who described the condition contemporaneously with Dr. Robert Graves. In my research on toxic stress, I had come across some data describing the high number of cases of hyperthyroidism among refugees from Nazi prison camps. In fact, the term *kriegs-Basedow* (*kriegs* means "war," so *kriegs-Basedow* is "hyperthyroidism of war") was coined following the observation of an increased incidence of hyperthyroidism during major wars. Trinity visited the endocrinologist, who confirmed that she did in fact have Graves' disease. Undoubtedly, her hyperthyroidism was contributing to her issues in school. Once Trinity was on medication, her behavior and learning problems improved. They weren't gone, but she was doing a heck of a lot better than she had been before.

It turns out that since 1825, researchers have known that Graves' disease is often correlated with stressful life events, which Trinity had in spades. It was clear that her problems with emotional regulation were overlaid on the hyperthyroidism, making her time in the classroom that much more difficult. The crazy thing is that many busy physicians do their entire assessment of ADHD based on behavioral symptoms alone, without a stethoscope even touching the patient's chest.

Once again, I saw how critical it was to take a whole-system approach to examining kids who were at high risk. Even if I didn't always know exactly what I was looking for, using the ACE score as a measure of *risk* for toxic stress was making me a better doctor, helping me put the right lens on the problem so I could detect things I might otherwise overlook. After prescribing medication to treat Trinity's Graves' disease, which was the first story her symptoms were telling me, I prescribed family therapy to treat the second story her symptoms were pointing to—underlying toxic stress. The purpose of family therapy was to teach Trinity and her aunt how to create an environment that would limit the reactivation of her SAM and HPA axes. The goal was to give them the tools to prevent scary or stressful situations and to manage them better when they came up, essentially reducing Trinity's dose of adrenaline and cortisol.

I didn't start Trinity on any medications for her behavior; I favor a stepwise approach to treating toxic stress so I can see what's working and what isn't. There are certainly some patients for whom medications are an important part of treatment, but our clinical team is careful to use medications in a way that addresses the underlying biology. In the previous chapter, I mentioned that a graph of the response of the prefrontal cortex to adrenaline and noradrenaline looks like an inverted U. Well, for kids with impaired impulse control and inattentiveness due to toxic stress, PFC function is likely to be on the downslope of the inverted U (kind of like if you drink *way* too much coffee, you can't focus to save your life). In those cases, our clinical team tends not to use stimulants like methylphenidate (Ritalin) or drugs derived from amphetamines. Instead, we often use guanfacine, a nonstimulant that was originally developed to treat high blood pressure but has also been used to treat ADHD. Guanfacine targets specific circuits in the prefrontal cortex where adrenaline and noradrenaline exert their action, improving impulsiveness and concentration, even in situations of high stress.

While I felt good about taking a more systemic approach, like the doctors who first began to suspect that a compromised immune system was behind HIV/AIDS, I was working on a medical frontier. There wasn't (and still isn't) a clear set of diagnostic criteria or a blood test for toxic stress, and there is no drug cocktail to prescribe. My biggest guide for what symptoms might be toxic stress–related was the ACE Study itself, but I knew that the number of diseases and conditions it accounted for might just be the tip of the iceberg. After all, if a dysregulated stress-response system was the source of the problem, it could have far-reaching effects. A disrupted stress response doesn't affect only the neurological system, it affects the immune system, the hormonal system, and the cardiovascular system as well. Because everyone's biological and genetic makeup is different, how that dysregulation manifests itself will be similarly diverse.

Right about here is where my staff started to get overwhelmed with what we were learning, feeling as though *everything* could be toxic stress–related. When we talked through it, I reminded them that it was all about where you started with the problem. If you broke it down, the core issue was a dysregulated stress response. From there you simply followed the thread, looking at how that dysregulation affected each of the body's systems. We made a choice to start our investigations with the underlying systems. If we wanted to identify and treat what was wrong, we had to know what was happening on a molecular level. We turned back to the literature and tried to break it down system by system, figuring out as best we could exactly *how* toxic stress was disrupting the normal functions of the body.

Toxic Stress and the Brain

Based on the results of our chart review, it seemed that learning was the proverbial canary in the coal mine. The fact that our patients with four or more ACEs were 32.6 times as likely to have been diagnosed with learning and behavioral problems signaled to us that ACEs had an outsize effect on children's rapidly developing brains. I had learned a lot about brain development in medical school and residency. I understood that a child's brain forms more than one million neural connections every second during the first years of life. I'd also seen firsthand during my medical residency that if that process got disrupted, by a toxin, a disease, or even physical trauma, the consequences could be serious.

Now we needed to understand the many ways that toxic stress affected the brain. The science nerd in me liked to think me and my team were akin to the rebel army in the movie *Star Wars,* searching the plans of the Death Star, but in this case, the Death Star was toxic stress. If we knew how the Death Star worked, studied its blueprints, looked for its weaknesses, we might find a way to prevent the harm it could cause.

• • •

In the previous chapter we talked about the cast of characters in the stress response: the amygdala, the prefrontal cortex, the hippocampus, and the noradrenergic nucleus of the locus coeruleus (which we'll refer to as the locus coeruleus from now on). Because these parts of the brain are on front lines of the stress response, it makes sense that a severe and prolonged disruption of the norm would hit them hardest, changing how they fundamentally do their jobs. Another very important region of the brain in understanding how ACEs create long-term problems is the ventral tegmental area (VTA). This is the pleasure and reward center of the brain and it plays a huge role in behavior and addiction.

The Alarm (aka the Amygdala)

The amygdala is the brain's fear center. It's located deep inside the temporal lobe near the midline and is believed to be one of the first brain structures that evolved, which is why it's often referred to as the "lizard brain." The amygdala is a key player in a series of interconnected parts of the brain that together make up the limbic system, which governs emotions, memory, motivation, and behavior. The amygdala is one of the most important structures in the limbic system because it helps you identify and react to threats in your environment. Fear is an emotion that developed to help you save your skin from the bear and that erupts when you first hear a roar or catch a glimpse of the animal's hulking profile.

When the amygdala is repeatedly triggered by chronic stressors, it becomes overactive, and what we see is an exaggerated response to a stimulus like the bear or, as I was beginning to notice in clinic, a nurse with a needle. MRI studies of severely maltreated kids from Romanian orphanages shows dramatic enlargement of their amygdalae. The other thing that happens when the amygdala is chronically or repeatedly activated is that it starts messing up its predictions about what's scary and what's not. The amygdala begins sending false alarms to the other parts of your brain about things that shouldn't actually be scary, just like the little boy who cried wolf.

I Don't Know Karate but I Do Know C-razy (aka the Locus Coeruleus)

This part of the brain is the driving force behind aggressive behavior (sorry, Raiders fans, I'm still looking at you). It works closely with the prefrontal cortex, which is why we see overlap in how they both regulate impulse control. The dysregulated locus coeruleus releases too much noradrenaline (the brain's version of adrenaline) and can result in increased anxiety, arousal, and aggression. It can also seriously mess with your sleep-wake cycles by overloading your system with hormones that tell it to remain vigilant because (hello!) a bear is in your cave.

The Conductor (aka the Prefrontal Cortex)

The prefrontal cortex (PFC) sits right behind your forehead at the front part of the brain. Unlike the amygdala, which is thought to be a very primitive structure, the PFC is believed to be one of the last to have evolved, and it confers faculties of reason, judgment, planning, and decision-making. It is often referred to as the seat of "executive functioning," which is the ability to differentiate among conflicting thoughts and inputs, consider future consequences of current activities, work toward a defined goal, and exhibit "social control" (that is, suppress urges that, if not suppressed, lead to socially unacceptable outcomes). In many ways, it's like the conductor of an orchestra, setting the tempo and volume for each of the different players, harmonizing all their inputs into something that is coherent and beautiful, not chaotic and loud. Think about your average day in a fifth-grade classroom. The teacher is talking, the kid beside you is throwing a wad of paper across the room, your archnemesis is kicking you ferociously under the table, and the girl you like just passed you a note telling you that she doesn't like you anymore. This is a lot for a *normally* functioning PFC to deal with.

For kids with toxic stress, the activity of the prefrontal cortex is inhibited in two ways. First, the overactive amygdala sends messages to the PFC telling it to decrease its functioning because something scary is happening; you don't want reason getting in the way of survival. The second is that the locus coeruleus is flooding the brain with noradrenaline, compromising the ability to override instincts and impulses. The PFC is the part of the kid's brain that puts the brakes on impulses and helps him or her make smarter decisions. Telling a kid to sit still, concentrate, and ignore stimuli that are flooding his brain with the *need to act* is a lot to ask. This down-regulation of the PFC can have different consequences for different people. For some, it results in an inability to concentrate and solve problems, but in others it manifests as impulsive behavior and aggression.

Memory Bank (aka the Hippocampus)

The hippocampi are two cute little seahorse-shaped parts of the brain responsible for creating and maintaining memory. When the amygdala gets activated during a major stress event, it sends signals to the hippocampus that disrupt its ability to knit together neurons, essentially making it more difficult for the brain to create both short-term and long-term memories. On brain scans of patients with Alzheimer's disease, the hippocampi are badly damaged. Knowing that, it's pretty obvious why this part of the brain is so critical to learning, and it's easy to see how kids with quick-trigger amygdalae are behind the eight ball when it comes to everything from memorizing multiplication tables to spatial memory.

Vegas, Baby!
(aka the Ventral Tegmental Area, VTA)

If the locus coeruleus is a Raiders fan, then the VTA region of the brain is Las Vegas. Responsible for things like rewards, motivation, and addiction, this part of your brain is the one you really don't want running away with your credit card. Basically it all boils down to dopamine, which is the feel-good (or feel-*amazing*) neurotransmitter that peppers your brain with rewards when you have sex, shoot heroin, or say yes to that piece of triple chocolate cake at the end of the day.

When your body's stress-response system is overloaded again and again, it messes with the sensitivity of your dopamine receptors. You need more and more of the good stuff to feel the same amount of pleasure. The biological changes in the VTA that lead people to crave dopamine stimulators like high-sugar, high-fat foods also lead to an increase in risky behavior. The ACE Study shows that there is a dose-response relationship between ACE exposure and engaging in many activities and substances that activate the VTA. A person with four or more ACEs is two and a half times as likely to smoke, five and a half times as likely to be dependent on alcohol, and ten times as likely to use intravenous drugs as a person with zero ACEs. So for anyone looking to prevent young folks from developing dependencies on bad-for-you dopamine stimulators like cigarettes and alcohol, understanding that exposure to early adversity affects the way dopamine functions in the brain is an absolute must.

Hormonal Harmony

Ladies, have you ever noticed that the *one* month you are sweating about whether you are going to get your period is the month that it seems to come late? Well, it's not just your imagination. Due to the impact it has on the hormonal systems, the stress response can affect everything from menstrual cycles to libidos to waistlines.

Hormones are the body's chemical messengers, responsible for kicking off a wide range of biological processes. Big ones include growth, metabolism (how your body gets and stores energy from food), sexual function, and reproduction. So, basically, everything. The hormonal system is very sensitive to the stress response. Which makes sense, because when you see the bear in the woods, it's hormones that get the party started *("Adrenaline! Cortisol! Go!")*.

Just about every one of the body's hormonal systems is affected by stress. Growth hormones, sex hormones (including estrogen and testosterone), thyroid hormone, and insulin (which regulates blood sugar) all tend to decrease during stress. Some of the major health impacts are dysfunction of the ovaries and testes (also known as gonads), psychosocial short stature, and obesity. In the case of gonadal dysfunction, for women this can lead to not ovulating, not having a period, or menstrual irregularity. In one study, researchers found that 33 percent of newly incarcerated women with stress (can you imagine a newly incarcerated woman who *doesn't* have stress?) had irregular periods. Psychosocial short stature is what we saw with Diego — severe delay of growth in children and adolescents due to a pathological environment. In some cases, children have severely reduced levels of growth hormone, but other times, as we saw with Diego, growth hormone isn't measurably decreased. In these cases, we believe the disruption comes from the other factors that help growth hormone do its job. Obesity is a much more familiar foe, but in the hormonal system, we see the double whammy. As I mentioned above, because of its impact on the pleasure center (the VTA), chronic stress increases your cravings for high-sugar, high-fat foods, and elevated cortisol makes it harder for your body to metabolize sugars and easier for your body to store fat. But cortisol isn't the only bad guy here; the hormones leptin and ghrelin are also increased with activation of the stress response. Together they intensify appetite and work with cortisol to do their worst for your waistline.

• • •

The chart review that we did at the clinic showed us that if a kid had an ACE score of four or more, he or she was twice as likely to be overweight or obese as a child with zero ACEs. This is where we see how biology and social determinants of health collide with significant consequences. We've talked about how kids living in vulnerable communities have a lot of intersecting risks driving ill health. Lack of access to good health care, few safe places to play, and food insecurity do contribute to striking health disparities in places like Bayview.

But our patients with zero ACEs lived in the same neighborhood and had the same access to health care, the same lack of safe places to play and nutritious food as our patients with high ACEs. When you realize what toxic stress does to the hormonal systems of kids who have experienced multiple ACEs, you understand that it's not *just* because they subsist primarily on a diet of fast food that they are overweight. It's not *just* that they are living in a food desert (a term that refers specifically to a neighborhood with a dearth of nutritious food) and are being brought up by parents who think Taco Bell is a healthy alternative to McDonald's. Those things compound the problem, to be sure, but they are not the whole story. Our data suggested how powerful the underlying mechanism of toxic stress can be — that the metabolic disruption was also an important driver. If you grow up in a food desert, of course it's going to be difficult for you to be healthy. But if you *also* have higher cortisol levels that are driving you to crave high-sugar, high-fat foods, it's going to be that much harder for you to choose broccoli over French fries.

Foreign Relations: Toxic Stress and the Immune System

Immunology was by far my most painful class in medical school, which is ironic considering that the immune system should be the doctor's best friend. The problem is the intricacy of it all. The immune system wields a lot of power; it is responsible for monitoring the relationships between what's inside and what's outside in the world and also for defending the body against foreign threats. Kind of like your own personal secretary of state and secretary of defense rolled into one. Because the body has so many different antagonists and so many different allies, sometimes it's hard to tell them apart. The immune system has to be an expert on *all* of it, knowing, for instance, that the protein on the outside of a bacteria or virus is *bad* and the microbe needs to be fought off, but also that the proteins in the lungs, nerves, and blood cells are *good* and should be left alone.

When the body's secretaries are pleased with foreign relations, they are very low-key. They quietly go about the business of maintaining order by constantly scanning the body for cells that are infected, injured, or becoming cancerous, and when they find them, they destroy them. But when a bad guy manages to evade the routine defenses and cause disease, the secretary of defense sounds the alarm, marshaling armies and launching strategic attacks. The immune system uses chemical signals called cytokines to activate your body's response to injury or illness. The word *cytokine* literally means "cell movers." They prod your body to make more white blood cells, which fight off infection and activate different types of cells to do things like make antibodies and eat bacteria. The immune system also stimulates inflammation (like when a bug bite gets all red and swollen). Like everything else in the body, what's important in the immune system is balance.

Dysregulation of the stress response has a profound impact on immune and inflammatory responses because virtually all the components of the immune system are influenced by stress hormones. Chronic exposure to stress hormones can suppress the immune system in some ways and activate it in others, and unfortunately none of it's good. Stress can lead to deficiency in the part of the immune system that fights off the common cold, tuberculosis, and certain tumors. In Sweden, researcher Jerker Karlén and his colleagues found that kids with three or more exposures to early stress showed increases in cortisol levels and were more likely to be affected by common childhood health issues such as upper respiratory infections (colds), gastroenteritis (stomach flu), and other viral infections. We also know that dysregulation of the stress response can lead to increased inflammation, hypersensitivity (think allergies, eczema, and asthma), and even autoimmune disease (when the immune system attacks the body itself), as with Trinity's Graves' disease.

In the years since the ACE Study was first released, scientists have looked closely at the relationship between ACEs and autoimmune disease. Research findings show a strong correlation between childhood stress and autoimmune disease in both children and adults. In partnership with Dr. Felitti and Dr. Anda, researcher Shanta Dube analyzed the data of over fifteen thousand ACE Study participants, looking at their ACE scores and how often they were hospitalized for autoimmune diseases such as rheumatoid arthritis, lupus, type 1 diabetes, celiac disease, and idiopathic pulmonary fibrosis. What Dube found was striking: a person with an ACE score of two or more had twice the odds of hospitalization for autoimmune disease as someone with zero ACEs.

Just as the brain or the nervous system is not fully developed when a child is born, the immune system is also still developing well after birth. In fact, when babies are first born, they have very little function-ing immunity, something that will develop with time and a little help from their moms. Breastfeeding is so important in part because the mom's antibodies protect the baby from infection and help grow his immune system. If you've ever wondered why people are hesitant to bring very young babies out into the world, that's why. (Well, that and the soul-crushing sleep deprivation.)

A baby's immune development occurs in response to his or her en-vironment over the course of the first years of life. Think about it as the secretary of state in her first year in office, still meeting all the foreign heads of state, feeling out who is hostile and who is friendly. Unfortu-nately, getting a good read on the reality of the threat is difficult when there is an adrenaline and cortisol overload. This kind of disruption early on in development can lead to lifelong alterations in the func-tion of the immune system and, in many cases, to disease. Think of it like this: If the secretary of defense is triggered to send in the troops to fight invaders in the body, sometimes the troops will attack the right enemies, but sometimes they'll find trouble where trouble doesn't ex-ist. The more inflammation there is in the body, the greater the chance that some of that inflammation will attack the body's own tissues, leading to autoimmune diseases like rheumatoid arthritis, inflamma-tory bowel disease, and multiple sclerosis. Because early adversity in-creases inflammation, when you have higher numbers of troops roam-ing around the body, there is a greater likelihood that they'll make a mistake.

Researchers in Dunedin, New Zealand, demonstrated that the changes in levels of inflammation were actually measurable. They fol-lowed a group of a thousand people over the course of thirty years, observing and recording a number of important health data points over that time. In addition to reinforcing Felitti and Anda's findings, the Dunedin researchers discovered that even twenty years after their subjects had been maltreated as children, four different markers of in-flammation were higher than they were in those who hadn't been mal-treated. What makes this study a critical addition to the research on ACEs is that the patients' adverse childhood events had been reported *as they were happening,* strengthening the case for causality by documenting that the adversity preceded the biological harms.

We know that a well-balanced immune system is critical to good health. When we realize that adversity in childhood harms the development and regulation of the immune system *throughout someone's life,* we begin to understand just how powerful the ACE science can be to combat some of the leading causes of disease and death.

• • •

For me, the immune-system piece of the ACEs puzzle was important because I found that when people learned how toxic stress affects the immune system, they listened in a different way. It's counter to the story they may already have in their heads. People seem to know that if you eat too much, you mess with your hormones and gain weight, and if you make impulsive decisions or become addicted to alcohol, you'll affect your neurological system. But it's harder to connect those perceived human failings to something like Graves' disease or multiple sclerosis. Most people don't think about those conditions as being caused by anything other than genetic bad luck. What is so powerful about the follow-up ACE studies like the one Dube did is that they show a strong correlation between autoimmune diseases and exposure to something environmental and specific — childhood adversity.

Dr. Felitti's patient Patty is a perfect example of why it's important to pay attention to those correlations. Patty was extremely obese and also had some psychological and emotional problems (the sleep-eating is the tip-off there). Even for those who know that abuse often leads to emotional problems and sometimes to obesity, those issues might seem like the beginning and end of the impact of adversity on her life. But when we see that Patty actually died of idiopathic pulmonary fibrosis, an autoimmune disease (the odds of which increase with the number of ACEs a person has), the plot thickens. The consequences of toxic stress are not just neurologic and hormonal; they are also immunologic, and those symptoms are much more difficult to spot. Patty's childhood adversity threatened her immune system as much as it did her mental well-being. The problem was that, for Patty, no one suspected that her immune system could be fatally compromised because of toxic stress. No one knew where to look.

. . .

My understanding of how early adversity affected my patients had come farther in the past twelve months than it had in the previous decade, yet the picture wasn't quite complete. It made sense to me that an overactive stress response could do a lot of harm to someone's health. I felt I understood clearly how the changes to the neuro-endocrine-immune systems could lead to problems for my kids. But the ACE Study also showed that adversity in childhood could lead to health problems decades later. By that time, many people would have escaped the challenging conditions of their childhood. So why was Dr. Felitti seeing the same or, arguably, worse problems in his patients? How was it that ACEs were the gift that kept on giving? I had the niggling sense that the blueprint of the toxic stress Death Star went one dimension deeper, drawn in even fainter lines. I knew these questions would take me even farther down the rabbit hole of toxic stress, but I had come this far and I had to find out how it worked on the deepest level of all: genetics.

6

Lick Your Pups!

PARENTS OF VERY YOUNG babies come into my clinic displaying all the different colors of the emotional rainbow — exhausted, elated, concerned, proud, terrified. So when Charlene brought her daughter, Nia, in to see me, Charlene's complete lack of facial expression stood out. When I asked this young mother a question about her daughter, she would respond, but her face and eyes stayed flat. It was almost as if we were talking about what size shoe she wore or what time the number 22 bus would arrive. Otherwise, she could have been any early-twenty-something mom with an infant; she was snugly tucked into jeans and wore a cute blouse with her hair pulled back neatly. Five-month-old Nia, however, was not typical — when Charlene was pregnant with her, Nia had stopped growing and had to be delivered by emergency C-section eight weeks early; at birth, she had clocked in at a mere three pounds. After weeks in the hospital, Nia improved nicely and was released in good health, but in the weeks that followed at home, she had struggled to gain weight.

As I worked with my team and Charlene to figure out the cause, I became increasingly concerned. We spent hours walking Charlene through how to prepare food for her daughter, when to feed her, and how much to give her. We took Nia's vitals and did blood work. We watched her weight and height measurements like mission control for the space shuttle launch. All the while, Charlene, too, was coming into clearer focus. Moving beyond her characteristic flatness, she would quickly become annoyed and overwhelmed when her daughter would cry or fuss. She'd tell her to shut up or ignore her completely. It looked to me like a clear case of postpartum depression, but no amount of urging could convince Charlene to get help.

Eventually, Nia's health became critical and we were out of options. She was suffering from failure to thrive, a medical term that describes babies who don't gain enough weight and eventually can't meet their developmental milestones. Every second in the first years of life over one million new neural connections are formed, so if an infant isn't getting enough fats and proteins needed to make healthy brain connections, that can have significant impacts. I recommended that Nia be hospitalized, hoping that under constant care, she would gain the weight she so desperately needed. Nia spent four days in the hospital and did exactly that, but soon after she was released, the gains she made were erased. We redoubled our efforts, bringing in our social worker and trying hard to engage Charlene in treatment, but eventually we had to send Nia for yet another hospital stay. This time, when I talked to the inpatient team at the hospital, we agreed that it was time to start talking about Child Protective Services (CPS). They were seeing the same problems with Charlene and Nia's dynamic that we were. Charlene was still suffering from depression, and she was still refusing to get help. After Nia was released the second time, she again failed to grow and thrive at home. With a heavy heart, knowing Charlene would be thrown into a tailspin, I had to do something no pediatrician ever wants to do—file a CPS report.

I didn't know for sure that Charlene was being overtly neglectful, not feeding Nia, or hurting her, but I did know that Nia was way below the third percentile for weight even when we took into account her prematurity. She was in the danger zone and it was clear by then that the dynamic between daughter and mom was affecting Nia's growth. In cases like this, it can be hard to parse things out. We know that babies who are born premature are at greater risk of neglect simply because they have greater needs—more irregular sleep patterns, more frequent feedings—and that those needs can be enough to overstress an exhausted new parent. But if an infant doesn't have a caregiver's reciprocal eye contact, stimulating facial expressions, snuggles, and kisses, hormonal and neurologic damage can occur, and that can prevent a child from growing and developing normally. When a baby is not being cared for, she doesn't grow well, even if she has enough nutrition. Was Nia's problem that she wasn't getting enough food? Or was it that Charlene was so depressed she wasn't stimulating Nia? The truth was that it could have been both.

Here's where I put my toxic stress lens on the situation. At the tender age of five months, with a depressed mom and a dad who wasn't involved, Nia already had two ACEs. I had some strong suspicions that Charlene had an ACE score as well. Despite the initial sadness I felt at having to file the report and put Charlene under the strict eye of CPS, a major question I'd had before came bubbling to the surface once again: How is it that ACEs are handed down so reliably from generation to generation? For many families, it seemed that toxic stress was more consistently transmitted from parent to child than any genetic disease I had seen.

Take, for example, Cora, a longtime Bayview resident who was the primary caregiver for ten-year-old Tiny, her great-grandson. At sixty-eight, Cora had not intended to raise another child, but when the child welfare workers called to say that Tiny's mom was incarcerated and they needed to find a home for the boy, Cora felt torn. Her son, Tiny's grandfather, wasn't capable of caring for a child. Both he and Tiny's grandmother had struggled with addictions to alcohol and other substances, and she had passed away from kidney failure in her late forties. Now it looked like Tiny's mother would be in prison for a long time. Cora was exhausted. Still, she couldn't let the boy go into the system.

Cora brought Tiny in to see me for his regular checkup. Her greatest concern was his behavior. She received calls from the school on a daily basis. Most recently, he had overturned his desk in class, and when the teacher pulled him aside to reprimand him, Tiny had kicked her, earning him a suspension. During the exam, I got a chance to see what Cora was talking about. Most kids are on their best behavior in the doctor's office, so observing Tiny was revealing. He would frequently interrupt, aggressively rip up the exam table's paper to get our attention, and then leap off the table, open drawers, and pull out whatever was inside. At one point, he scooted down on the floor and managed to unplug my computer before I could redirect him. No doubt, staying ahead of Tiny was a workout.

Cora and Tiny's visit was in the early days of the Bayview clinic, back before we were doing regular ACE screening, but I could tell he would need a lot of resources. I excused myself for a moment to knock on Dr. Clarke's door for a brief consultation. When I returned to the room, I opened the door the same way I always did, with a brief "Knock, knock" before I gently swung the door open. The scene I walked in on stopped me in my tracks.

Tiny crouched in the corner, his hands shielding his face from the blows his great-grandmother was raining down on him. Shoulders, head, face, body — Cora was slapping and yelling, really going at him.

I almost couldn't believe my eyes. Was she seriously beating the child *in the doctor's office?*

"*Stop!*" I said forcefully. I crossed the room in two strides and physically inserted myself between them. "You're not allowed to hit children in our clinic or anywhere else."

I gave Tiny a good once-over to make sure that he wasn't seriously injured. Then I calmly explained to Cora that because I was a mandated reporter, I would have to call CPS.

"Go on an' call 'em," Cora responded. "CPS don't got to raise that baby, I do. He need to get some act-right in him. Otherwise, he goin' to end up in the pen just like his mama."

It was obvious to me that Cora believed that she was doing the right thing. After watching two generations lose their way, Cora was relying on the tools she had learned in her own upbringing to keep Tiny on the straight and narrow. The irony was that, despite Cora's intentions, the beating was undoubtedly unleashing a neurochemical cascade that made Tiny *more* likely to end up like his mom and his grandparents. That day, I convinced Cora to sit with me as I made the call to CPS. She got to see that I wasn't "ratting her out" but rather advocating for her, telling the agency that she needed additional tools to help her manage Tiny's challenging behavior without using violence. Ultimately, she trusted me enough to agree to work with Dr. Clarke; the beatings stopped, and the family remained intact.

. . .

For a long time, that interaction with Cora stayed with me. I thought about her and Tiny and the generations in between. I was seeing all around me evidence of multigenerational ACEs. But it was rat mothers and rat pups in landmark studies by Dr. Michael Meaney and his colleagues at McGill University that helped me piece together how to understand and ultimately interrupt the biological legacy of toxic stress.

Meaney and his team looked at two groups of rat mothers and rat pups. They noticed that after the pups were handled by researchers, the moms would soothe their stressed-out pups by licking and grooming them. This is basically the human equivalent of hugs and kisses. What was fascinating was that not all moms did it to the same extent. Some moms exhibited high levels of licking and grooming behavior toward their pups. Other moms displayed low licking and grooming behavior, which meant they didn't give as many sloppy kisses and embarrassing hugs when their pups were having a rough day.

Here's the part that made me sit up straight in my chair: Researchers observed that the development of the pups' response to stress was directly affected by whether the mom was a "high licker" or a "low licker." They found that pups of high-licker moms had lower levels of stress hormones, including corticosterone, when they were handled by researchers or otherwise stressed out. This high-licker-leads-to-low-stress effect also showed a dose-response pattern: the more licking and grooming the rat pups got, the lower their levels of stress hormones. In addition, the pups of high-licker moms had a more sensitive and effective "stress thermostat." By contrast, pups of low lickers not only had higher spikes of corticosterone in response to a stressor (in this case, being placed in restraints for twenty minutes), they also had a harder time shutting off their stress response than did the pups of high-licker moms. The licking and grooming behavior that occurred in the pups' first ten days of life predicted changes to their stress response that lasted for the entire *lifetime*. Even more startling, the changes continued into the *next generation*, because female pups who had high-licker moms became high lickers themselves when they had their own kids.

I thought of Charlene and Nia as I read about Meaney's work and I wondered how much "licking and grooming" Charlene herself had received as a child. She was certainly facing her fair share of stressors. I had witnessed in my residency how frightening it can be to have a premature infant, even for the most well-supported and resilient of parents. When she came through the door of my clinic, Charlene was the young, depressed mother of a premature infant, but she hadn't always been that.

Growing up in Bayview, Charlene was full of promise. As a high-school soccer star, she seemed to have beaten the odds when her athletic prowess earned her a college scholarship. But a knee injury in her freshman year cut her dreams short. She dropped out the following year, and after a few years at home, she became pregnant. Now she was struggling to care for her baby girl. I worried for both Charlene and Nia. My medical training had taught me how to make the diagnosis of failure to thrive. What I hadn't learned was how to break the intergenerational cycle of toxic stress.

I consumed Meaney's research, searching for that all-important mechanism at the source. What the researchers were hoping to discover was *how* this early behavior could go on to affect the rats' stress response and behavior for the rest of their lives. In other words, these scientists were looking for the root of the change. Just like me.

What they found was that the rat moms were, in fact, handing down a message to their pups that changed the way the pups' stress responses were wired, but the mechanism, the *how* of the changes, turned out to be not genetic, but *epigenetic*.

Many people still think of genes and the environment as very separate things: you're born with a certain genetic code that determines your biology and health and you have experiences that shape more malleable things like character and values. Keeping genes and environment in separate corners like this has sparked years of debate about which is more important, nature or nurture. People have been arguing over this for a long time, but as science gets more and more advanced, there is less and less to argue about. Scientists can now say pretty definitively that there is no separating the two. In fact, we now know that *both* environment and genetic code shape *both* biology and behavior. Considering how closely genes and environment work together, it's no surprise that the debate raged on for hundreds of years with no winner in sight. Luckily, with the advances in science, we are finally able to see that there is a vital synchronicity that determines what we look like, how our bodies work, and ultimately *who we are.*

Most people know that DNA is the genetic code, the basic blueprint for your biology. To take that understanding a step further, your body uses this code as a template to produce the proteins that make up new cells and ensure that all the things inside those cells function. Every cell has your entire genetic code in it as well as the machinery to read the code and decide which parts of the sequence to translate into proteins.

Environment and experience play a huge role in determining which parts of your genetic code are read and transcribed in each new cell your body creates. How does your experience or environment do that? Well, it turns out that the body doesn't actually "read" every "word" of its DNA. What scientists have discovered is that baked into the cells are both the genome (your entire genetic code) and the epigenome, another layer of chemical markers that sit on top of your DNA and determine which genes get read and transcribed into proteins and which ones don't. The term *epigenetic* actually means "above the genome." These epigenetic markers are handed down from parent to child along with the DNA.

One way to think about it is this: The genome is like the musical notes in sheet music and the epigenetic markers are like the notations that tell you how loudly, quietly, quickly, or slowly to play the notes. There might be a notation to skip an entire section of music altogether. These epigenetic notations are subject to experience, to being rewritten by your environment.

Activation of the stress response is one big way the environment can change epigenetic notations. As your body tries to adapt to the stress of your experiences, it turns certain genes on or off, particularly genes that regulate how you'll respond to stressful events *in the future.* That process of the epigenome working with the genome to respond to your environment is called *epigenetic regulation* and it's critical to our understanding of why toxic stress is so damaging to our *lifelong* health. When a four-year-old breaks a bone, that trauma is not encoded in his epigenome; it doesn't affect him in the long term. But when a four-year-old experiences chronic stress and adversity, some genes that regulate how the brain, immune system, and hormonal systems respond to stress get turned on and others get turned off, and unless there is some intervention, they'll stay that way, changing the way the child's body works and, in some cases, leading to disease and early death.

There are a handful of processes that are responsible for epigenetic regulation, but the two that we know the most about when it comes to the genetics of stress are DNA methylation and histone modification. In DNA methylation, a biochemical marker called a methyl group is attached to the beginning of a DNA sequence. That marker prevents the gene from being turned on; it acts like a Do Not Disturb sign hanging on a hotel doorknob. It tells the DNA housekeeping team not to come in and translate that genetic sequence into proteins, essentially rendering that part of the genetic code silent.

Histones are like a chastity belt for the DNA. They are proteins that keep the genetic material locked up, preventing the DNA transcription machinery from getting to it. When certain biochemical markers are attached to the histones, the histones are then modified — they change shape and become more open, allowing the DNA to be read and transcribed. Which brings us back to the rat moms and their pups. The "lick your pups" study is a great example of this type of epigenetic regulation. Meaney and his team found that high-licker moms were releasing high levels of serotonin in their offspring. You may have heard that serotonin is the body's natural antidepressant. It boosts mood and acts as the equivalent of rat-pup Prozac. This serotonin didn't just make the pups *feel* better, it also activated a chemical process that changed the transcription of the part of the DNA that regulates the stress response. Meaney and colleagues eventually demonstrated that all that licking and grooming ultimately changed the epigenetic markers on the rat pups' DNA, leading to lifelong changes in the stress response.

This kind of epigenetic change is like a communication shortcut for nature. When rat moms don't lick their pups, they are essentially telling them that there is something to be wary about in the environment, so they should be on high alert. Instead of waiting around for the generations-long process of genetic adaptation to change the offspring's

DNA, this environmental information gets passed on to the rat pup quickly through a change in the epigenome. To look more closely at this process, the Meaney research team did something brilliant; taking a cue from a Lifetime TV movie, they switched some of the rat pups at birth. They placed pups of high-licker moms with moms who were low lickers, and vice versa. The study found that the pups' DNA methylation took on the pattern of their foster moms', *not* their genetic moms'. So did their behavior — if a rat pup born to a high-licker mom was fostered by a low licker, she grew up to be an anxious adult rat with high levels of stress hormones who was a low licker herself when she had her own pups. Meany and his team found that the differences in licking and grooming that happened very early on (in this case, the first ten days of a rat pup's life) made a huge difference.

To take it one step further, Meaney and his colleagues tested whether it was possible to reverse DNA methylation patterns *after* a rat had reached adulthood. Using trichostatin A, a solution capable of pulling methyl markers off DNA, they devised a way to *chemically* alter methylation patterns. When they injected the TSA solution into the brains of the adult offspring of both the high-licker and low-licker moms, it completely eliminated the changes in the adult rats' stress response.

This study was a showstopper for me for a couple reasons. It showed the mechanism of these long-term changes was not simply genetic. The adverse experiences of my Bayview patients were factors that extended down to their DNA and likely changed them *epigenetically.*

Meaney's work showed me not only how moms can negatively affect their pups by not licking them enough but also how they can *help* them by licking them more. The fact that environment is something we can modify means there is a lot of hope for human pups born to "low-licker" moms. These pups are not damaged goods; they are not defective. If they can get a safe, stable, and nurturing environment at an early age, the biology says that this sets them up to develop a healthy stress-response system in adulthood. As we've mentioned, the key to keeping a tolerable stress response from tipping over into the toxic stress zone is the presence of a buffering adult to adequately mitigate the impact of the stressor. In the case of the rat pups, it's the mom's licking and grooming. In the case of a human, it could be a dad hugging and listening. The buffer is hugely important, not just to attenuate the stress hormones but also to prevent the kind of epigenetic changes that lead to a dysregulated stress response and the major health issues that come with it.

. . .

But I still had some questions. We know that a rat pup whose mom was a low licker would likely have lifelong problems with the regulation of its stress response. And we also know that an overactive stress response can set off a cascade of changes to neurologic, endocrine, and immune function. But on the level of DNA, how does that chronic stress affect the likelihood of getting certain diseases, like cancer? After looking at how changes to the epigenome can be passed down from generation to generation, I wondered if higher risks for particular diseases became embedded as well. Was there some part of the DNA that got changed by stress and permanently turned on the genes for disease? Or was there something else going on? It wasn't until I stumbled into the wild world of telomeres that I saw there was more than one way to reprogram DNA.

. . .

It'll probably be no surprise that the only thing I love more than a badass scientist is a badass *woman* scientist. So you can imagine my excitement when I found out about a dynamic duo right in my own backyard. I was first introduced to the work of Dr. Elizabeth Blackburn and Dr. Elissa Epel of UCSF by a friend who has many lovely qualities but who also happens to be a *little* obsessed with premature aging. When it comes to aging, I tend to ignore the chatter and stick with clean livin' and night cream, but when my friend dropped the words *chromosomes* and *premature cell death* into the conversation about the latest antiaging news item, my ears perked up. Turns out this was one *legit* scientific discovery in the quest to understand the aging process. Dr. Blackburn is one of three scientists who received the Nobel Prize for discovering how telomeres, the sequences on the ends of chromosomes, work to protect DNA from the kind of damage that can lead to premature aging and death. Blackburn teamed up with health psychologist Elissa Epel and the two took off on a research tear, exploring how exactly telomeres could be shortened or damaged and, more important, how to stop it.

Blackburn and Epel looked at how food, exercise, and even mental focus affected the health of telomeres. But to me, the most interesting part of what they found was that *stress* had a major impact on the length and health of telomeres, and that in turn had a major impact on the risk of disease.

Let's back up a second. So what exactly are telomeres again? Sequences? It always helps me to think of telomeres as the bumpers at the ends of DNA strands. Telomeres are *noncoding* sequences that, for a long time, no one thought much about. They don't make proteins and at first glance aren't super-active in the body. But researchers discovered that they actually do serve a vital purpose: telomeres protect DNA strands, making sure that every time it is replicated by cells, the copy is true to the original. Telomeres are very sensitive to the environment, which means that, like good car bumpers, they always take the first hit. Anything biochemically harmful (like stress) is going to damage the telomeres much more than the DNA. When the telomeres are hurt, they send signals to the rest of the cell letting it know that the bumpers have taken too many hits and that the cell should respond. The cell reacts in two major ways. The first is that when the telomeres get too short (too many bad parallel-parkers in the neighborhood), the cell can become senescent, which is a science-y word for old. This means the cell retires and doesn't do its job anymore. Take collagen (the protein in skin that makes it supple and prevents wrinkles). If too many of the fibroblast cells that are supposed to be making collagen hit the road to play shuffleboard at Del Boca Vista, you're left looking a decade older than you actually are.

Lots of things can damage the telomeres and lead to premature cellular aging, but chronic stress is a big one. When a cell becomes old or dies, it's not the end of the world, but if there is too much cell death in one place, it can lead to poor health. For instance, if there is too much cell death in the pancreas, you won't be able to make enough insulin, which can lead to diabetes. The response a cell can have to damaged and shortened telomeres other than senescence is that it can become precancerous or cancerous. When that happens, it means the ability of the cell to copy its DNA correctly has been compromised, and it begins coding for mutations that say, "Keep making cells forever!" This causes the cells to replicate uncontrollably and grow into a tumor that continues to grow and grow and grow. Simply put, if there is too much damage to your telomeres and they become excessively shortened, it can lead to premature cellular aging and disease or cancer. This adds yet another fun variable to the dating game; not too far into the future, ladies might start looking for partners with long telomeres.

Research on telomeres and stress is relatively new, but we do know that early childhood adversity predicts shorter telomeres in adults, showing us the lasting imprint that early stress has on cellular aging and disease processes. Elissa Epel worked with researcher Eli Puterman and other colleagues to examine data for 4,598 men and women collected as part of the U.S. Health and Retirement Study. They assessed cumulative adversity for both childhood and adulthood by reviewing responses to health questionnaires. For childhood stressors, criteria included a participant's family receiving help from relatives because of financial difficulties, the family relocating due to financial difficulties, a participant's father losing his job, a parent's substance abuse or alcohol use causing problems in the home, whether the respondent had experienced physical abuse before age eighteen, repeated a school year, or gotten in trouble with the law. The questions about adult stressors surveyed death of a spouse, death of a child, qualifying for Medicaid, experiencing a natural disaster, being wounded in combat, having a partner addicted to drugs or alcohol, being a victim of a physical attack, or having a spouse or child with a serious illness. Epel and Puterman then looked at each respondent's telomere length. They found that while lifetime cumulative adversity significantly predicted telomere shortening, that shortening was due mostly to the adversity experienced in childhood; adult adversity on its own was not significantly associated with telomere shortening. For each childhood adversity a study participant experienced, his or her odds of having short telomeres increased by 11 percent. Epel and Puterman's data also showed that household adversities, such as abuse or having a parent who used alcohol or drugs, were a stronger predictor of telomere shortening than household financial stress.

Further work by researchers Aoife O'Donovan and Thomas Neylan compared the telomeres of people with PTSD with the telomeres of people in good mental health. What they found was that overall, those with PTSD had shorter telomeres than those in the control group. However, what was really interesting was that the people with PTSD who *did not* have early childhood adversity *didn't* tend to have shorter telomeres.

The good news is that even if you have shortened telomeres, maintaining healthy telomeres can protect you from further shortening. How do you keep your telomeres healthy? One important way is by boosting levels of telomerase, which is an enzyme that can actually lengthen the telomere. Once again, the science is new, but it suggests that even if you start out with shorter-than-normal telomeres, you can still slow decline by increasing your telomerase with things like meditation and exercise.

• • •

So does that mean genes don't matter? All you need is a mom who licks and grooms you a lot? Not so fast. While the epigenetic part of the equation is new and exciting and tells us a lot we didn't know, there's no discounting the impact of the DNA that comes from the good old egg and sperm. As we know, it's all about nature *and* nurture. You are handing down to your kids both your genome *and* your epigenome and they both count in determining health. For instance, you might be blessed with some crazy-long telomeres. Maybe every woman on your mother's side of the family has lived to be over a hundred while never looking a day over seventy-five. But during early childhood you experienced adversity, and now you have a high ACE score. Your telomeres are being chipped away at faster-than-normal rates, but because of your genetically long telomeres, you've got a cushion. In that case, there may not be a dramatic result; you're not necessarily going to live to be a hundred, but you also might not see the premature mortality that your ACE score would predict. However, if you don't have the genetic advantage of long telomeres, it could be a different story. If you go through childhood adversity, the shortening of your telomeres could lead to worse health outcomes than you might otherwise experience. And just like two siblings with the same parents might have different eye colors, they also might have different lengths of telomeres, which can lead to different outcomes even if they experience similar doses of adversity.

· · ·

The research on epigenetic regulation and telomeres reinforced what I already suspected—early detection is critical. Now more than ever, I believed if we could identify those at risk for toxic stress by screening for ACEs, we had a better chance of both catching related illnesses early and treating them more effectively. Not only that, but we could also possibly prevent future illness by treating the underlying problem—a damaged stress-response system. If we put the right protocols into place in pediatric offices across the city, country, and world, we could intervene in time to walk back epigenetic damage and change long-term health outcomes for the roughly 67 percent of the population with ACEs *and* their children. And, someday, their great-grand-children.

The potential for outcomes like these and the science behind them had me fired up. I had already graduated from talking people's ears off at cocktail parties to reaching out to every well-connected person I knew in the medical community in search of someone who had more power than I did and who would commit to doing something. My own clinic had already begun instituting routine ACE screening for every patient, but there were so many other doctors out there who could benefit from this information. Having grown up in Palo Alto in the eighties back when it was closer to middle class (as opposed to straight-up wealthy like it is now), I knew that kids with ACEs live in lots of different kinds of neighborhoods. Several of my classmates at a Palo Alto high school attempted suicide when I was there, and I later heard stories of parental substance abuse and mental illness the students had struggled with in secret. Even in areas much better off than Bayview, toxic stress was essentially invisible to the health-care system.

Bayview might be a fairly obvious place to look for the impact of adversity, but toxic stress is an unseen epidemic affecting every single community. Since the original ACE Study was published, thirty-nine states and the District of Columbia have collected population ACE data. Those reporting their data show that between 55 and 62 percent of the population have experienced at least one category of ACE, and between 13 and 17 percent of the population have an ACE score of four or more. The states with the highest rates of ACEs among young children were Alabama, Indiana, Kentucky, Michigan, Mississippi, Montana, Oklahoma, and West Virginia. Left unchecked, the effects of ACEs and the toxic stress they create were being handed down by well-meaning parents in families all across the country and, undoubtedly, around the world.

After a great conversation with Dr. Martin Brotman, the CEO of California Pacific Medical Center at the time and my stalwart champion, I saw my chance. Every hospital CEO in San Francisco was part of an organization called the Hospital Council of Northern and Central California. This group came together for lots of reasons, but one of its many jobs was addressing health-care disparities in the city. Dr. Brotman helped lead the health-disparities task force within the council and was excited about what I had told him about ACEs and our work at the clinic. He immediately invited me to give a presentation about ACEs to the council. Feeling the kind of excitement that almost makes you want to throw up, I left his office that day thinking, *This is it!* This was my chance to go to the decision-makers and the health-care-shapers and blow the lid off this thing. I'd better not screw it up.

I spent weeks preparing for my presentation.

On the day, I knew I was ready, but as I sat in the lobby after showing up ridiculously early, I realized I hadn't ever been this nervous, not even for my medical boards. I had just a small block of time on the CEOs' agenda, and when I was finally shown into the room, they were all there. Mostly older men, mostly white, there were roughly twelve of them, comfortably spread out around a U-shaped table, papers stacked and strewn around their salad plates, multiple beverages stationed next to laptops. Some smiled pleasantly while others nodded. For a minute I cursed my bad luck for having gotten a slot at the end of what was obviously a very long business meeting. If I couldn't keep them riveted, I hoped I could at least keep them awake. Dr. Brotman stood up and graciously introduced me. I shook hands with everyone and then made my way to the front of the room and popped my jump drive into the computer. After what felt like the longest thirty seconds of my life, the drive connected and I pulled up my first PowerPoint slide.

I looked up and noticed a short, heavyset Caucasian woman in the back silently clearing plates and refilling coffee. It crossed my mind briefly that I wouldn't mind trading places with her. A tremor of self-doubt unsettled me for a moment. I took a deep breath. If this were about me, I wouldn't even be here. No way. But this was for my patients. With that in mind, I silently exhaled and started talking. For a good twenty-five minutes I held forth, trotting out the data, the science, the biological mechanisms. Like Dr. Felitti, I was convinced that once people saw the figures, the sheer numbers of people living with the effects of ACEs, they would be blown away. I didn't talk about my patients at all; I talked about their stress-response systems. Months of practicing my talking points at borderline socially inappropriate moments had helped me polish what I thought were my most powerful arguments.

Finally, I stopped.

I paused for a few moments, hoping to let the import of it all sink in. Then I said some approximation of "Okay, guys, so what are you going to do about it?"

I looked at their expressions, and I could tell immediately that their reaction was not going to be what I had hoped. My stomach tightened. A slow-moving burn began to make its way across my face, spreading the embarrassment cell by cell. My body may have registered it before my mind, but rapidly I knew one thing. Though it seemed they all agreed that what I had just said was both striking and important, they could clearly tell that I was profoundly naive about how things worked. What was written in their expressions was soon followed up by statements amounting to something along the lines of "Okay, Nadine, what are *you* going to do about it?"

Looking back on it, I realize that all I did was present them with a problem. When they asked me questions about solutions, I didn't have good answers. They probed me about screening protocols and wanted to know what best-treatment practices were and how I thought they could be implemented. I tried my best to explain that right now there wasn't a protocol for anything. That was why I was coming to *them*. Wouldn't they figure out how to implement the best universal screening tools and come up with protocols for other doctors? That was their job, right?

Judging by the thrust of their questions, it sure wasn't.

It became pretty clear that the CEOs weren't going to take up this cause on their own time, despite the fact that they were supportive of it. In terms of priority, it certainly wasn't going to jump the line ahead of seismic upgrades for their buildings or the next audit from the Joint Commission on Accreditation of Healthcare Organizations. How naive was I to think that they would just drop everything for this? I sputtered through my goodbyes, all the while feeling like a cartoon balloon, slowly and sadly deflating in the middle of the room. I don't really remember how that meeting ended, what I said, or who sent me on my way with a kind nod and a handshake. There is still a bit of a fog about the last couple of minutes of the meeting.

Eventually, I reached the elevator and proceeded to repeatedly jam my finger against the Down button.

I had worked really hard, I had prepared, I had convinced them, but still nothing was going to come of any of this. I had been living in the world of ACEs and toxic stress so intensely for so long that it felt like the most important thing in the universe. It was straight-up weird to me that I could explain this to other doctors and they could see it too and even agree but still not jump out of their chairs. I wasn't mad or upset at them—I was just confused. My sense of confidence in reality as I knew it was shaken, and this led me to a line of questioning I hadn't entertained before. What if this puzzle I had put together about adversity wasn't the five-alarm fire I thought it was? Even worse, what if there wasn't anything we could do about it?

III

Prescription

7

The ACE Antidote

LEAVING THE HOSPITAL COUNCIL meeting that day, I was so distracted by my self-defeating questions that I didn't even notice when she first called out.

The elevator yawned open.

"Excuse me, Doctor?" she repeated.

I turned and saw that it was the woman who had been pouring coffee for the CEOs in the conference room at the beginning of my presentation.

"Yes?"

She took a tentative step toward me. Up close, I could see that she had a rough-looking dye job and one tooth missing on the right side, but she was neatly tucked in and buttoned up in her hotel uniform. I paused for a moment and then let the elevator door close behind me, giving her my full attention.

"That's me," the woman said.

"Pardon?"

"That's me that you were talking about up there. Those ACEs — the bad things that happen to people when they're kids — all of that stuff you were talking about has happened to me. I've got every single one of those. I think I'm a ten out of ten."

She paused and took a deep breath, shifting her gaze down to a small, dark gray tattoo on her left wrist.

"I've been working to stay sober and I've had lots of problems with my health. After hearing what you had to say just now, I feel like I finally understand what's been going on with me."

Her eyes met mine. "Anyways, I just wanted to say . . . thank you. Keep doing what you're doing."

"What's your name?" I asked.

"Marjorie," she said, smiling.

I smiled back.

"Thank you, Marjorie."

. . .

Since that day with Marjorie and the hospital council, after every talk and every presentation, I make it a point to go up to the people clearing the tables or breaking down the PA system to ask them what they thought. No matter how well my presentations are received professionally, talking to these folks always gives me additional insight into how the story of ACEs is playing out in people's day-to-day lives. I walk away understanding that no matter the geographies, ethnicities, and socioeconomic backgrounds, we are all affected by ACEs in similar ways. I was trained to believe in the power of clinical medicine and public health to improve lives, yet it is clear from these conversations that many people who have experienced ACEs and are grappling with their lifelong effects don't know what they are dealing with. No doctor has ever told them that there might be a problem with their stress-response system, much less suggested what to do about it. Those few minutes in front of the elevator with Marjorie served as both a touchstone and a swift kick in the butt. If we didn't have a clinical protocol to address ACEs and its many health impacts, then it was time to *create* one. Fortunately, I was too naive to understand how huge a task that would ultimately be.

On a small scale, we were already making progress at the clinic, so I knew we were on the right track. Along with screening all children for ACEs at their annual checkups, we were actively putting the toxic stress lens on our treatment plans and starting to look for evidence-based treatment models that focused on the underlying biology of children, parents, and communities dealing with the impacts of adversity. Outside of ours, there were no pediatric clinics I knew of that routinely screened for ACEs in 2008. Patients with toxic stress were most likely to come to the attention of their pediatrician with symptoms of behavioral problems or ADHD, which, as it turns out, was good news for them, because it meant that they were likely to be referred to a professional in the mental-health field, one of the few health-care specialties that had recognized the link between early adversity and poor health. Unfortunately, many physicians had no clear understanding that clinical illnesses like asthma and diabetes might also be manifestations of toxic stress. As we saw with Diego, psychotherapy was in fact one of the most well-supported therapeutic interventions for patients with symptoms of toxic stress whether those symptoms were behavioral or not.

When primary-care doctors have easy access to mental-health services for their patients, those patients have a better shot at getting the treatment they need. To that end, one of the best approaches for helping doctors who care for patients with ACEs and toxic stress (which, statistically speaking, is every single doctor in America) is integrated behavioral health services. That simply means having mental-health services available at the pediatrician's (or primary-care clinician's) office. Later I would find out that this was an emerging best practice, one now being endorsed by just about every national health-care oversight agency, including the U.S. Department of Health and Human Services. The Bayview community had asked for mental-health services before I'd read the ACE Study—that's why I brought Dr. Clarke onboard. Having a mental-health clinician in our office was so successful and Dr. Clarke was in such high demand that I was soon looking for more mental-health resources to pour into our clinic.

For most pediatricians working in low-income, underserved areas, like I was, the available resources would typically be limited to a referral to a community agency, possibly a social worker if you were lucky, and then you'd cross your fingers and maybe say a few prayers. But in the months leading up to when I started treating Nia, we had begun working with Dr. Alicia Lieberman at the University of California, San Francisco, a renowned child psychologist who specialized in child-parent psychotherapy (CPP). This type of therapy focuses on children from birth to five years old and is built on the notion that to help young kids experiencing adversity, you have to treat the parent and child like a team. The groundbreaking aspect of CPP, and what Dr. Lieberman believes makes it so effective, is the recognition that real conversations with kids about how trauma is affecting them and their families — even when kids are really little — are critical.

Alicia Lieberman recalls, as one of her earliest memories, the experience of waking up in the middle of the night to an odd feeling of movement. Growing up in Paraguay during a time of political revolution and unrest, she saw her father, a pediatrician who spoke out about the social injustices he witnessed, become a target of the government. He was periodically jailed for interrogation, but as a respected member of the community, he was returned each time. The growing civil unrest left the family constantly on edge. More and more community leaders were being jailed or simply "disappeared."

When Alicia awoke that night, she saw that her mother and father were carrying the bed with her still in it. Her parents were transporting their sleeping daughter to the innermost room of the house to protect her from stray bullets that might come through the walls. Eventually, she and her family emigrated, taking a transatlantic ocean liner to Israel. On the ship, a fellow traveler asked the young girl what it was like to live under that kind of stress. At the mention of the events they were leaving behind, Dr. Lieberman remembers tensing up and having the realization that *stress lives in the body.*

Dr. Lieberman started her professional work from a place of deep personal familiarity and curiosity about trauma and stress. On top of the instability and fear of the family's political circumstances, when Alicia was four years old the tragic death of a sibling threw her parents into a state of profound grief. The surviving children were never told what happened and young Alicia was left to create her own narrative, a story conjured up by her imagination out of confusion and sadness. As she got further into the study of child psychology, she saw that talking about the past openly and honestly with children wasn't common practice. The thinking at the time was that little children didn't understand things like death and violence and if you tried to talk to them about it, you would just retraumatize them. Dr. Lieberman doubted that the practice of telling Santa Claus stories to children when bad things happened was doing them any good.

Dr. Lieberman debunked the long-held myth that young children and babies don't need treatment for trauma because they somehow don't understand or remember the chaotic experiences they faced. Her work is built on research that shows that early adversity often has an outsize effect on infants and young children, just like it did on Dr. Hayes's tadpoles. After years as a clinician, Dr. Lieberman came to understand that children's need to create a story or narrative out of confusing events is actually very normal. Children are compelled to give meaning to what is happening to them. When there is no clear explanation, they make one up; the intersection of trauma and the developmentally appropriate egocentrism of childhood often leads a little kid to think, *I made it happen.*

Dr. Lieberman sought to explore ways in which both parents and children could talk openly and honestly about trauma. She also rightly recognized that parents' own rough childhoods and the scars that they still carried might affect the way they responded to their child in stressful or traumatic circumstances, hindering their ability to act as a protective buffer. She learned from her mentor Selma Fraiberg that families can learn how to "speak the unspeakable" and that parents can discover tools to support and buffer their children, even in moments of crisis. Eventually, Dr. Lieberman would go on to codify the CPP protocol and demonstrate its efficacy in five separate randomized trials. Supported by the latest science, CPP is now one of the country's leading trauma treatments for young children, and it is instrumental in helping the whole family begin to heal.

CPP takes into consideration all the other pressures and drama that both parent and child have to deal with — other family members, the community, work (or lack thereof) — everything that affects the parent-child bond. This allows patients to make connections between the traumas of the past and the stressors of the present, so they can better recognize their triggers and manage their symptoms.

Traditionally, if a mom is depressed, she finds her own therapist and they work one on one. CPP's approach is based on the understanding that the quality of the relationship and the health of the *attachment* between the parent and child are absolutely fundamental to health and well-being. There was hardly a clearer case of this than Charlene and

Nia. Fortunately, Dr. Todd Renschler, a postdoctoral fellow under Dr. Lieberman's supervision, was just joining our team when Charlene and Nia first came into my waiting room. Charlene was understandably furious with me for months after I filed the report with Child Protective Services, but in their case it was exactly what needed to happen. In order to keep custody of Nia, Charlene was required to get help with her postpartum depression, which meant intensive psychotherapy.

When Charlene came to her first CPP session with Dr. Renschler, she had her iPod earbuds dug in deep with the volume turned up so loud he could have tapped along to the beat. She plopped Nia down on the couch beside her and stared blankly at Dr. Renschler. Needless to say, the first sessions were pretty challenging. Charlene felt betrayed by me and felt she was being forced to do something against her will. An experienced and patient clinician, Dr. Renschler took his time building rapport with Charlene, starting off by giving her some choice in how the sessions would proceed, offering her some power in a situation where she felt totally powerless. Instead of diving right into Nia's health and Charlene's depression, he started by addressing what Charlene said was her biggest problem, something that every parent of an infant can relate to: serious lack of sleep. Nia was waking up frequently in the night and Charlene was exhausted and frustrated.

It was no surprise that Charlene and Nia were struggling with sleep. Researchers have found that infants of depressed moms have a harder time regulating their sleep; they sleep an average of ninety-seven fewer minutes a night than infants of nondepressed moms and have more nighttime awakenings. Childhood adversity significantly increases the risk for just about every sleep disorder there is, including nightmares, insomnia, narcolepsy, sleepwalking, and psychiatric sleep disorders (sleep-eating, anyone?). Nighttime sleep plays a powerful role in influencing brain function, hormones, the immune system, and even the transcription of DNA.

Sleep helps properly regulate both the HPA and the SAM axes. During sleep, levels of cortisol, adrenaline, and noradrenaline drop. As a result, lack of sleep is associated with increased levels of stress hormones and increased stress reactivity. As you know from Chapters 5 and 6, these stress hormones kick off the party, triggering brain, hormone, immune, and epigenetic responses to stress. The downstream effects are impaired cognitive function, memory, and mood regulation.

Sleep deprivation doesn't just make you groggy and cranky; it also makes you sick. Lack of sleep is associated with increased inflammation and reduced effectiveness of the immune system. While you're catching z's, your immune system does a systems upgrade, using the downtime to calibrate its defenses. Everyone knows it's important to get sleep when you're sick, but it's just as important when you're healthy. Lack of sleep leaves people more susceptible to illness because the immune system doesn't appropriately fight off the viruses and bacteria that it is constantly exposed to.

Poor sleep is also associated with reductions in hormones such as growth hormone and with changes to DNA transcription, which for children can be especially problematic, opening the door to issues with growth and development.

Dr. Renschler worked with Charlene to create a routine that would help Nia sleep for longer stretches. He started by helping Charlene understand the importance of putting Nia to bed in a cool, dark, and quiet environment at the same time every night, avoiding stressful or stimulating activities just before sleep and instead giving her a soothing bath and reading a story before bedtime. Eventually both mom and baby started getting some much-needed shuteye. Feeling understood and ultimately supported in this problem helped Charlene believe Dr. Renschler knew what he was doing. More important, she saw that he was there to help *her*.

Soon, Charlene began to open up about the lack of support she had. Her ex-boyfriend (Nia's dad) had been abusive during her pregnancy and was now out of the picture. She lived with her maternal aunt, who had raised Charlene and her little brother since their mother committed suicide, when Charlene was a young child. Ever since she had told her aunt she was pregnant, she'd received more criticism than support. Despite living with her aunt, she felt completely isolated, and it only got worse when Nia was born so prematurely. The further Dr. Renschler and Charlene got in their conversations about her relationship with her aunt, the more she expressed wanting to have a different kind of relationship with Nia. In a nuts-and-bolts way, meeting this goal came down to examining how she was interacting with Nia. In the CPP sessions, when Nia cried or smiled, Dr. Renschler encouraged Charlene to think about how that felt and what she thought it meant. Once, when Nia was in her lap, the baby reached up and pulled out Charlene's earbuds. At first she was annoyed with her daughter's "bad behavior," but when Dr. Renschler wondered aloud about what else Nia could be communicating with that action, Charlene admitted that maybe her baby just wanted her attention. Charlene's aunt was critical, distant, and unwilling to give her the kind of support she was craving, so when similar dynamics seemed to be playing out with Nia, Dr. Renschler helped Charlene recognize that and think about how she might respond differently.

Soon, the relationship started to shift. Charlene began taking one earbud out during sessions and, finally, both. As she became more tuned in to her daughter, Nia responded with fewer cries and more of the more coos and laughs that, as any parent knows, are the sweet rewards that make up for all the midnight feedings and cranky mornings. Charlene also began to take a more active role in solving her baby's failure to gain weight. In her sessions with Dr. Renschler, she wanted him to help her fix the bottle at just the right temperature and asked a lot of questions about baby food and feeding. Our clinic team worked together to support Charlene with practical advice, nutrition information, and access to resources. We also regularly communicated as a team about Nia's progress. Through these supportive conversations, Charlene's resentment about the CPS report began to fade, and she became less angry with me.

While Charlene was doing great in her therapy sessions and her relationship with Nia, she continued to have issues with her aunt. One day, she made baby food for Nia (a big step for her!) and forgot to put away a bowl after she was done. Her aunt was so pissed off that she told Charlene she could no longer use the kitchen. Charlene felt frustrated and defeated. There she was, trying to do the right thing, and her aunt was punishing her for a small oversight. But the incident opened up space for Charlene to talk more with Dr. Renschler about her relationship to her aunt, the loss of her mother, and even her feelings of help-lessness and depression following Nia's birth. Her aunt had been angry when Charlene got pregnant, and without her aunt as a support system, Charlene had felt completely alone. Then the baby had stopped growing suddenly and had to be delivered via emergency C-section, and no one could tell Charlene why. After all, she wasn't smoking or taking drugs, and as far as she knew, she had been doing everything right. At the time, we didn't have any answers for her. It wasn't until later that I learned just how closely ACEs and high doses of maternal stress were related to premature birth, low birth weight, and increased rates of miscarriage.

When Nia was in the NICU, Charlene was completely physically disconnected from her child. Nia didn't look like any baby Charlene had ever seen before. She was small and frail with multiple tubes and monitors connected to her tiny body. Charlene was terrified that her daughter would die and she began to wall herself off emotionally. People leaving was something Charlene was accustomed to. She had never known her father, and her mother had left her and her brother when Charlene was just five years old. In a way, Charlene was preparing herself for the inevitable — the loss of her daughter.

Through her conversations with Dr. Renschler, Charlene realized that it was actually possible to talk through some of these difficult experiences. She wished that she could do this with her aunt. But her aunt, who had lost a child as a young mother, had her own wall up, making the intergenerational cycle of distance, disconnection, and stress seem impenetrable. As Dr. Renschler and Charlene worked together over time, Charlene began seeking a replacement for that maternal connection. Though her ex, Tony, was out of the picture, his older sister was welcoming of Charlene and wanted to have a relationship with Nia. Charlene started taking her daughter over to see her paternal aunt and began spending more and more time there. Dr. Renschler explained to Charlene that forming caring relationships, like the one she now had with Tony's sister, was an important ingredient for health, both her child's and her own.

Then, seemingly out of nowhere, Charlene stopped coming to therapy. Dr. Renschler didn't see her for two weeks, and though he phoned and left several voicemails, his calls were never returned. When she finally came back, Charlene had the faint outline of a black eye, and her earbuds were firmly in place. A crying Nia sat beside her on the couch and Charlene was once again staring blankly at the wall. All those months of progress seemed to have evaporated. Only gradually did Dr. Renschler get the full story from Charlene. She had been visiting Tony's sister with Nia when Tony showed up out of the blue, agitated and ranting. While she was holding Nia, he suddenly attacked her. Terrified, she ran away to call the police, leaving Nia with Tony's sister. Following the attack, it was as if Charlene and her daughter had been transported back in time. Nia was up all night, screaming and inconsolable, and they were whisked back to the land of no sleep. Over the next several sessions, it became clear that what had happened with Tony had sent Charlene back into a depression and Nia into a state of distress. During one session when Nia was crying inconsolably, Charlene said to Dr. Renschler, "She just gets so mad at me." They talked more about how Charlene felt when Nia screamed and cried, and Charlene admitted to worrying that Nia was going to be short-tempered like Tony. She got mad at Nia for crying because she didn't want people to think her ten-month-old baby was crazy like her dad.

. . .

Charlene kept going to CPP and she and Dr. Renschler worked hard to find a path back to the success they'd had early on. During a particularly hard session, Charlene quietly put her hand on her stomach. When Dr. Renschler asked what she was experiencing, she explained this was what she did when she was really upset, something that helped her calm down when she felt she was going to lose it. Dr. Renschler told her it was actually a really good sign that she could recognize when she felt that way. Often when people's stress response becomes activated, their biological systems are so overstimulated that they don't know what to make of it. This lack of understanding means that people don't take time to collect themselves; they just react in whatever ways their bodies tell them to — lashing out at others, acting impulsively, or self-medicating. For Charlene, this made intuitive sense.

The conversation about biology opened the door for Dr. Renschler to discuss mindfulness, the practice of being aware of internal thoughts and feelings in a sustained way. There were several calming techniques that Charlene could use when she was feeling stressed or overwhelmed, and she and Dr. Renschler worked on using breathing and awareness to focus and soothe her body's response to stress. Charlene started employing mindfulness strategies at home when she and her aunt fought and found it to be a big help. While the trauma with Tony definitely set Charlene back, eventually, after filing charges against him for assault and working through the shame and anger she felt about it, things got better. Dr. Renschler, with the support of the clinic staff, continued to work with Charlene and Nia on feeding, sleep, and mindfulness, reinforcing techniques that could be used again and again when things happened to trigger them both, bringing trauma to the surface.

The good news was that the healthier Charlene got, the healthier Nia got. Over time, she put on weight and caught up on her developmental milestones, and the CPS case was successfully resolved. Charlene started looking for work and even described to Dr. Renschler how she used her mindfulness exercises to help calm herself during a stressful job interview. She got the job, moved into her own apartment, and eventually got into a healthy relationship. By then, Charlene had forgiven me for the CPS report. I had made a point of checking in on mother and baby when they arrived to see Dr. Renschler. Eventually, we resumed our relationship for Nia's regular checkups. When Charlene came in and told me about getting the job, it felt like a victory. Instead of just treating the symptoms of Nia's failure to thrive, we had been able to treat the root of it — the stress caused by depression and trauma and an unhealthy family dynamic. Despite setbacks along the way, the child-parent psychotherapy had been a real success, changing the dynamic that was affecting Nia's health and strengthening Charlene's ability to act as a buffer for her child when problems arose.

To this day I will never forget the image of a chunky sixteen-month-old Nia toddling through the clinic, giggling and being chased by her mother. As a doctor, there are moments when you realize that you have saved a life. It's a tremendous feeling of satisfaction (mixed with exhaustion) that most often occurs in the chaos of the hospital after a successful resuscitation. As I saw Nia coming up the hall, I was struck with that same feeling: *We did good.*

. . .

As my colleagues and I made a conscious effort to look at our patients through the ACEs lens, the small victories started coming more and more steadily. While there were certainly challenges and stumbling blocks, we were having great success finding ways to help our patients with ACEs soothe their disrupted stress-response systems and manage their symptoms more effectively. We found that a focus on the underlying biology of toxic stress and the factors that helped balance the dysregulated pathways — sleep, integrated mental-health services, and healthy relationships — made a big difference for our patients. Soon, we were on the lookout for more tools to use in our toxic stress toolkit.

Pediatric obesity was one of the major health problems we targeted. With heartbreaking consistency, the 94124 zip code had the highest rate of obesity in all of San Francisco. Bayview is a food desert, which means there are way more fast-food outlets here than in other neighborhoods and almost nowhere to get fresh fruit and vegetables. I experienced this firsthand when I didn't have time to go food shopping and couldn't bring my lunch to work for a week. My options included all the greasy shades of fast food — taco truck, Taco Bell, McDonald's, KFC, and the least of the evils, Subway. Despite what its marketing department says, there are only so many days in a row a girl can eat a Subway sandwich.

Thanks to a grant from a local foundation, we were able to implement a cool obesity-treatment program modeled on a successful program at Stanford. Every Tuesday evening, two nutritionists from CPMC and two trainers from the Bayview YMCA came to the clinic to lead a group of our overweight patients and their parents. The kids went with the trainers to do some fun physical activity in a former warehouse space at the back of the clinic. It was a pretty bare-bones setup, but the area was large enough for a group of twenty kids to play volleyball, dance to Zumba, hula-hoop, and do whatever else would get them to work up a sweat. At the same time, their parents received hands-on instruction about how to prepare nutritious meals, and everyone ended the evening with a delicious, healthy dinner. To top it off, we had received some donated bicycles from a local company, so each kid who met his or her treatment goal would get a bike. You would think that this shiny kid-bait would be enough to keep my patients on track, but the truth was that most of our kids really struggled.

Bayview parents couldn't just let their kids run wild at the local park the way my parents did with my brothers and me. Parents in Bayview made sure their kids stayed safe by keeping them indoors — which meant that any stressful family dynamics were intensified. My colleagues and I knew that, as always, our kids with ACEs needed some extra help. To do that, we made sure that every patient in the program with a high ACE score (which was most of them) also received mental-health treatment with Dr. Clarke. Their therapy sessions focused on how their individual life experiences might be affecting their weight. The results were so good, it almost made me want to Zumba in celebration (almost). Pediatric obesity is a notoriously tough nut to crack, especially in communities like Bayview, but at the end of this program, every last bike was gone.

The program's success showed us that addressing ACEs as part of a weight-reduction program was essential. But in an interesting twist, we found that if our goal had been simply to address ACEs instead of obesity, exercise and nutrition would still have been an important part of that. It wasn't our initial intention to treat our patients' toxic stress with dodgeball and cooking classes, but we were pleasantly surprised to see how much the kids improved when we added healthy diet and exercise incentives to therapy. I sat down to check in with the moms and grandmas each week, and they reported that when they changed their children's diet and their levels of exercise went up, the kids slept better and felt healthier, and in many cases, their behavioral issues and sometimes their grades improved.

We found that there was plenty of science to support what we were seeing clinically. The data showed that regular exercise helped increase the release of a protein called BDNF (brain-derived neurotrophic factor), which basically acts like Miracle-Gro for brain and nerve cells. BDNF is active in parts of the brain important for learning and memory, like the hippocampus and the prefrontal cortex. We've long known that exercise improves cardiovascular health, but the research is piling up in exciting new directions, showing us that moving our bodies builds our brains as well as our muscles.

When it comes to combating toxic stress, addressing the dysregulated immune system is as important as supporting brain function. Regular exercise has also been shown to help regulate the stress response and reduce the presence of inflammatory cytokines. You might remember that cytokines are the chemical alarms that fire up your immune system and tell it to fight. For a person with toxic stress, moderate physical activity (like breaking a sweat for roughly an hour a day) can help the body better decide which fights to pick and which ones to walk away from. (While *moderate* exercise helps better regulate the stress response, there's no need to sign up for that ultramarathon. If you get too crazy, intense wear and tear on your body can actually increase cortisol levels.)

We saw that exercising made a huge difference for our kids, but so did eating right. Making a few specific changes to what grade of fuel went in the tank (e.g., substituting lean proteins and complex carbohydrates for greasy fast food) improved the body's ability to regulate itself. We explained that exercising and eating healthfully not only contributed to weight loss but also helped boost the immune system and improve brain function.

We've talked about how inflammation is one of the ways a well-regulated immune system fights infection, but as with everything else in the body, balance is critical. Too much inflammation causes all sorts of problems, from digestive issues to cardiovascular complications. Eating foods that are high in omega-3 fatty acids, antioxidants, and the fiber from fruits, vegetables, and whole grains helps fight inflammation and bring the immune system back into balance. By contrast, a diet high in refined sugar, starches, and saturated fats can promote further inflammation and imbalance. By choosing a healthier pattern of eating and adding moderate exercise to their routines, our patients had two great ways to bring their biological systems into better balance.

. . .

At that point, my staff and I had some strong strategies for specifically targeting and healing the dysregulated stress response: sleep, mental health, healthy relationships, exercise, and nutrition. Not surprisingly, these are the same things that, as Elizabeth Blackburn and Elissa Epel's research showed, boost levels of telomerase (the enzyme that helps to rebuild shortened telomeres). Of course, I was excited to find more. So once again, I pored through the literature looking for treatments that could lower cortisol levels, regulate the HPA axis, balance the immune system, and improve cognitive functioning. Over and over again the research pointed to one treatment in particular — meditation. Though many of us have been led to believe that meditation requires brightly colored robes and a mountaintop, or at least lots of crystals and green juice, training the mind has, fortunately, become a lot more mainstream than that. While techniques based on meditation practices began with religious sects thousands of years ago, they are now being used by an unlikely successor — the medical community. From cardiologists to oncologists, doctors have begun incorporating mind training into their clinical treatments.

Dr. John Zamarra and his colleagues looked closely at a group of adult patients in New York with coronary artery disease to see what (if any) effect meditation might have on their cardiovascular condition. Half of the group was randomly assigned to participate in an eight-month meditation program while the other half was assigned to a wait list. Everyone underwent a treadmill test at the start and end of the study. Remarkably, the biometric results demonstrated that at the end of the study, the patients in the meditation group were able to exercise on the treadmill 12 percent harder and 15 percent longer before experiencing chest pain. Even more interesting, during the treadmill test, the meditation group experienced an 18 percent delay in the onset of EKG changes that indicated stress on the heart, whereas the control group saw no changes to any of the clinical parameters. Researchers doing a similar study on meditation and cardiovascular health found a difference in arterial-wall thickness. Meditation was shown to be associated with reversing the narrowing of arteries, which for patients suffering from ischemic heart disease can be nothing short of lifesaving. In another study involving breast and prostate cancer patients, researchers found that meditation was associated with decreased stress symptoms, increased quality of life, and improved functioning of the HPA axis. Other studies have shown that meditation decreases cortisol levels, enhances healthy sleep, improves immune function, and decreases inflammation—all critical parts of keeping our biological systems balanced and able to mitigate the effects of toxic stress.

The more I read, the more it made sense to me. If stress can negatively affect the way the body works at a basic chemical level, then I could see how taking on a calming practice could positively change those same chemical reactions. While stress activates the fight-or-flight system (also called the sympathetic nervous system), meditation activates the resting-and-digesting system (also called the parasympathetic nervous system). The parasympathetic nervous system is responsible for things like lowering heart rate and blood pressure, and it directly counters the effects of the stress response. Given the profound connection between the stress response and the neurological, hormonal, and immune systems, a calmer, healthier mind seemed like a good place to start reversing the effects of toxic stress.

It wasn't long before I decided to take the science out of the journals and put it to work in the clinic. We quickly realized that reading the data on meditation was one thing but figuring out the right way to bring it to our patients was a whole different kettle of fish. I worried my patients would think meditation belonged in the hippie-dippie circles of the Haight-Ashbury district rather than in Bayview. What I really didn't want was a lady named Moonbeam coming in to tell my kids that they just needed to "find their center." I had to get my patients and their parents past the woo-woo factor and present meditation and mindfulness in a way that made them want to try it.

Being in the Bay Area, where cutting-edge science meets cultural sensitivity, I knew there had to be an in-between option; it was just a matter of time before I found it. And I did find it, in an impressive organization called the Mind Body Awareness (MBA) Project. MBA was doing mindfulness work (both meditation and yoga) with kids in juvenile hall and getting some solid results. I had seen the data on how many kids in juvie have their own fair share of ACEs (one study that came out later on looked at more than sixty thousand young people in the Florida juvenile justice system and found that 97 percent had experienced at least one ACE category and 52 percent four or more), so I figured it would be a good fit. After I met with MBA's executive director, Gabriel Kram, and heard his story, I was even more sold on our proposed partnership.

Gabriel grew up in an upper-middle-class home and attended an elite private high school in St. Louis, Missouri, before heading to Yale to study neurobiology. A few years in, he began a daily meditation practice, discovered how disconnected he felt from his authentic self, and dropped out of school. He passed through a period of intense anger and got caught up with a seriously shady crew. Never having been around people who didn't have his best interests at heart, Gabriel implicitly trusted them. One night, the leader of the group gave him a hit of LSD and then took him out with the intention of getting him to kill someone. He handed Gabriel a knife, identified the target, and shoved him toward the unsuspecting victim. Gabriel took a few steps and then paused. In that moment, a clear image of his father came to him. He realized that if he did this thing, he could never look at his father again without having to hide something. The image of his father literally stopped him in his tracks. That moment marked a turning point in Gabriel's life, and though traumatic, it opened a door to deep healing. When he later reenrolled in school, his mindfulness practice became the center that helped him stay connected to his values and integrity.

What motivated Gabriel's work with incarcerated youth was his realization that if it hadn't been for his father, for his stable and loving relationship with him, he might not have stopped himself from doing the unthinkable. And that love, that connection — it wasn't a given for every kid. Because of the possibility he had recognized in himself, he felt a strong desire to help those who didn't have a person like that in their lives, someone who stops you cold in a moment of truth. That safe, stable connection, along with the essential tools of mindfulness, had helped him immeasurably and he wanted to share that.

If you're ever lucky enough to meet Gabriel, the first thing you'll notice is his intensity. Far from being intimidating, he's totally magnetic, and as we sat down to plan our program I could already tell that my kids would love him.

To start out, we recruited fifteen girls with ACE scores of four or more for a ten-week program that involved a weekly two-hour session of mindfulness and yoga. I participated in the program with the girls and sprinkled in education about how the stress response works in the body, and how to recognize it and bring it back under control when it starts to go into overdrive. It was my favorite two hours of the week. The majority of my girls had experienced some type of sexual assault, and many of them had parents who were mentally ill or incarcerated, sometimes both. It was amazing to see the way the trainers from the MBA connected with our girls. By the end of the program, almost all of our girls reported feeling less stressed and, even better, as if they had new tools to manage stressful situations. Two of our girls stopped fighting in school, and most of them reported sleeping better as well as feeling more able to concentrate and connect in school.

With both our meditation program and our nutrition and exercise program, we saw the day-to-day evidence of progress, not by looking at numbers on a spreadsheet but by seeing individual kids literally dance into the waiting room, waving report cards that went from failing grades to honor roll. As their doctor, I got to see how, over time, they were hitting their clinical goals — better asthma management, weight loss, and so forth — but the special experiences for me were seeing Nia walk and Charlene smile and witnessing a kid with a sky-high ACE score lose ten pounds and take home a bike.

Slowly but surely, we were building our toolkit of clinical interventions to combat the effects of toxic stress. Sleep, mental health, healthy relationships, exercise, nutrition, and mindfulness — we saw in our patients that these six things were critical for healing. As important, the literature provided evidence of *why* these things were effective. Fundamentally, they all targeted the underlying biological mechanism —

a dysregulated stress-response system and the neurologic, endocrine, and immune disruptions that ensued.

I got to see all the ways these interventions were making my patients' lives better. I knew that was real, but as a scientist, I also knew it was anecdotal. We didn't have the manpower or the money to do the kind of systematic data tracking that would translate all those good report cards and bike-giveaway parties into solid research that would stand up to scrutiny in scientific circles. At one point, I even thought to myself, *We should be writing all of this up.* But our team was stretched thinner than pantyhose. I realized that we could do or we could write, but we didn't have the bandwidth for both. I decided that, for now, the doing was more important.

8

Stop the Massacre!

IN THE EARLY DAYS of the Bayview clinic, circa 2007, I was driving through the neighborhood when the car in front of me stopped suddenly.

At first, it was a mere annoyance. My mind was already thirty minutes into the future, engaging in a community meeting at the Bayview YMCA. About fifteen seconds passed before I realized it was time to swing the wheel left and go around. But just as I was about to make that move, a car coming from the other direction pulled up beside me and stopped.

A little alarm in my lizard brain started to go off. *What's going on here? This looks shady.* I checked the rearview mirror and got ready to jam the car into reverse, but before I could put my hand on the gearshift, another car wheeled around the corner and blocked me from behind.

I was trapped.

I could feel my body tense. With one hand on the steering wheel, I slowly reached for the automatic door locks. The guy in the first car got out and swaggered by me with a package. As he leaned forward to make the handoff to the guy in the car next to me, his shirt slid up to reveal the heel of a gun poking out from his waistband. *Holy crap!* My mind raced. *This is a drug deal! What if the deal goes bad and they start shooting? What if this guy sees me and decides I'm a witness?* My heart began to double-time it and my brain was like a radio locked on one station: How the Hell Do I Get Out of Here! I slouched down in my seat, willing myself to be invisible and, if possible, bulletproof.

Then, without so much as a look in my direction, the guy walked back to his car and drove away.

Minutes later, as I sat in my car unscathed, the radio station of my brain changed suddenly to the Holy Crap What Just Happened station.

After I finished freaking out, I immediately thought of my patients. On that day in 2007, I was still getting used to Bayview, but for my pediatric patients, this kind of thing could happen on their way to school or to the store any day of the week.

I learned early on that the threat of gun violence is a daily reality in Bayview, something you have to think about every time you walk to the corner store for a quart of milk. Years later, I met the district attorney of San Francisco, Kamala Harris, at a fundraiser right around the time we'd launched the mindfulness project at the Bayview clinic, and our conversation naturally turned to what we both saw as a devastating problem in a neighborhood we both loved. I'd heard Harris speak before, both on television and at events, and it was immediately obvious to me why people always talked about her as being the real deal, someone who was *getting things done*. She was young, charismatic, and knew how to energize a room. At first, I'd been a little hesitant to talk to her, but Harris was more approachable than I could have hoped, and pretty quickly my nervousness disappeared and we had a great conversation. She was curious about our work in Bayview and wanted to know more about toxic stress. It was refreshing to meet a politician who wasn't just delivering sound bites about how to make things better for people; I could tell she was actually *listening*. She seemed genuinely receptive to hearing different approaches to solving the community's problems.

When I started talking about Felitti and Anda's ACE Study, I discovered that Harris loved numbers as much as I did. She told me about an internal study she had done with the San Francisco Police Department. The department wanted to get a detailed look at the victims of homicide in the community, and one of the insights that emerged from that analysis had to do with the high rate of young murder victims. Among other things, the study found that 94 percent of murder victims under the age of twenty-five in San Francisco were school dropouts. As DA, Harris was the top prosecutor; her job was to be the official voice of the victims and go after perpetrators of crimes. But she wanted to know whether the city could find a way to *prevent* people from becoming victims of crime in the first place. What would that look like? She thought if she could devise a smart approach to stem the dropout tide, it would save lives. After all, kids who were in school weren't out on the streets, which meant they weren't victims of drive-by shootings.

Harris was interested in getting to the root of the problem, preventing rather than simply responding to the downstream effects once the chain of violence had been set in motion. Prevention is not something you hear DAs talk about every day, so when she told me about the re-direction program she was developing to keep kids in school, I was *seriously* impressed. I told her I thought she was right and that I believed we could go even further. I had recently heard a story about a pediatric emergency medicine doctor in Kansas City, Missouri, that seemed to point to the root of *both* of our problems.

. . .

Like Harris, Dr. Denise Dowd had been looking for ways to keep kids from getting shot. Her quest had started over a decade and a half earlier, in 1992, when a colleague of hers in the emergency department showed her an article in a local newspaper, the *Kansas City Star*. A journalist had profiled all the young people in the city who had died of gunshot wounds over the past year. The article included their photos and full names, and as the two doctors flipped through the profiles, they realized that a majority of the victims had been their patients. Many of the families used the ED as if it were their primary-care office, coming in any time their children needed to see a doctor. Over time, Dr. Dowd and her colleagues grew to know and develop relationships with their repeat customers. Now it was impossible not to wonder: Was there something they could have done? Could they find a way to recognize the next high-risk kid when he was sitting in front of them in the ED and help him before it was too late?

Dr. Dowd decided to do a chart review of all the pediatric firearm injuries in Kansas City for that year, looking for any factors that might be a common thread and possibly preventable. She obtained the health records, hospital admissions, EMS records, and coroner's reports of every child who had been killed by gun violence in the preceding year. What she found was that their medical histories revealed a pattern that repeated itself with tragic consistency. A typical story looked like this: A patient first comes in as a nine-month-old baby with a suspicious bruise, and the case is referred to Child Protective Services. The investigation is inconclusive. The next notation in his chart is from his pediatrician and details several missed visits for immunizations. At age four his preschool teacher complains that he won't sit still, has frequent tantrums, and hits other kids when he gets upset. He is diagnosed with ADHD and put on meds. At age ten, he's fighting and disruptive in school. This time, he's diagnosed with oppositional defiant disorder and put on more meds. At age fourteen he comes into the ED with a fracture of the fifth metacarpal, the bone in the hand that forms the knuckle of the pinky finger. Doctors call it a boxer's fracture because that's the bone that typically breaks when someone punches an object. The final entry in his medical record is at sixteen when he's brought into the ED for multiple gunshot wounds. This time he doesn't walk out.

. . .

In 2009, it seemed obvious to me that Dr. Dowd's prototypical patient was a clear example of untreated toxic stress. But in 1992, when Dr. Dowd was reviewing these charts, Felitti and Anda's research was still in the future. Dr. Dowd saw these similarities in medical history as a disturbing pattern, but the biological links had not yet been made.

After talking more about the ACE Study and other research on toxic stress, Harris and I agreed that we were looking at the same problem, just from different vantage points. I was trying to address kids' medical problems and she, like Dr. Dowd, was trying to keep kids safe. But what if we could put our heads together and address the potential root of *both* problems — ACEs. For the population of kids who were victims of gun violence, Dr. Dowd's research suggested we were likely to be dealing with a lot of high ACE scores. That meant a lack of impulse control and an impaired ability to focus — huge obstacles for getting kids successfully through school. For a kid with a dysregulated VTA (*Vegas, baby!*), pretty much anything from a trip to Taco Bell to smoking a bowl could easily win out over sitting in history class. How could we keep kids in school and safe *and* address the underlying biology that was putting kids at risk in the first place?

Harris and I continued our conversation about the profound social implications surrounding ACEs, health care, and the criminal justice system. One day I went to meet with her at the infamous Hall of Justice at 850 Bryant Street. (Anyone who has gotten a car towed in San Francisco knows that address all too well.) As we sat in her wood-paneled office, I shared with her some of my ideas that had coalesced since our initial meeting. I was convinced that if we could get more physicians like Dr. Dowd and myself to identify the kids in need of intervention early on, we could work to start healing their dysregulated stress responses so lifesaving programs like Harris's had an even better chance for success. We could prevent not only adverse health outcomes but also adverse social outcomes. I thought maybe she could use her position as DA to get the city to invest in research and data collection to find out if using the ACEs lens might make a difference.

Harris listened intently until I finished. Then she paused and looked me straight in the eye.

"Nadine, *you* need to be the one to make all these things happen. Start a center."

I laughed. "Girl, I've got my hands full doing what I'm doing."

"You and Victor could do it together. Think about it," she said in a kind, resolute voice that made it seem more like a foregone conclusion than a suggestion. She was the one who had introduced me to Victor Carrion and kindled the partnership that led to our chart review of patients at the clinic.

Harris would go on to become California's attorney general and then a senator, which gives you a good idea of how convincing she can be. I was flattered that she thought I could add the rigorous research and commitment to changing broad-scale awareness to the work that we were already doing, but I walked out that day thinking that she was vastly overestimating my abilities. She had the wrong woman. My experience starting the Bayview clinic, even with the full backing of one of the Bay Area's top-rated hospitals, was grueling. Long days, never enough money, fundraising, creating protocols, staff turnover — it felt like we had just gotten things working reasonably smoothly at the clinic. Starting an organization is really hard, and I wasn't in any hurry to do it again.

. . .

While a whole new center seemed out of reach, my discussion with Harris broadened my perspective. If ACEs were affecting not only health but social outcomes, I wasn't going to be able to work only the medical-community angle. I would need to talk to folks in education and criminal justice to learn more about how toxic stress related to the problems they were seeing.

The more people I met and spoke to about ACEs, the more I understood that the solution to this problem needed to be a lot bigger than the Bayview clinic. I knew from Dr. Felitti's data that 67 percent of Kaiser's middle-class, mostly Caucasian population had at least one ACE and that one in eight folks had four or more. It's one thing to read research papers that talk about prevalence rates and odds ratios. It's another thing entirely to meet the Marjories of the world and hear their stories. When statistics have faces, they feel a lot heavier. The worst part for me was thinking of the men, women, and children struggling with the effects of ACEs and toxic stress, walking around every day without knowing what the problem was and, harder still, not knowing that there were effective treatments. Their doctors don't tell them because chances are, their doctors don't know. To anyone looking at the day-to-day practice in the average doctor's office — or looking anywhere else in society — it was as if the research didn't exist. The more I knew, the more intolerable it felt to me that almost no one seemed to have this information.

As a result, I became even more vocal (if that's possible). Now when I went to medical and public-health conferences, I actively tried to influence the agenda to promote awareness about ACEs and toxic stress. As always, my work at the Bayview clinic both grounded me and continued to stoke the fire I felt around getting the word out. The only bad thing about coming home to Bayview was the reality of the clinic's minimal capacity for impact. There was so much urgency around *doing more* that conflicted with the small-potatoes nature of our operation. We had three exam rooms, one mental-health room, and one office. I shared that office with two other doctors and my research assistant Julia, which meant we couldn't all be in there at the same time. Dr. Renschler and Dr. Clarke were sharing the mental-health room, so we had to stagger their hours. The dentists who came from our partner clinic to deliver free dental services a couple of times a month set up "portable dental chairs" (I swear they looked like lawn chairs) and did dental screenings, cleanings, and fluoride applications in the warehouse space where we also locked up our charts and ran our exercise program.

In order to be able to answer the question from the hospital council and DA Harris — *What are* you *going to do about it?* — we would need researchers to help us measure the impact of our work. That was the only way we were going to be able to convince the hospital council, the city council, the *world* that there was something we could do medically about toxic stress. Dr. Carrion and his team could help us design studies that would stand up to academic scrutiny, but to do that work, they would need to be embedded in the clinic, and we literally did not have the room. We were like clowns in a clown car. At one point the concept of bunk desks crossed my mind. If we wanted to make a broad impact, we needed to test-drive treatments rigorously to ensure they would work in every pediatric office, not just ours.

Fortunately there was one person who often knew when I needed help before I did. Daniel Lurie was the founder and CEO of the Tipping Point Community, a grant-making organization that had a goal of ending poverty in the Bay Area. Tipping Point had been one of my biggest backers, helping us to launch the Bayview clinic and funding our partnership with Dr. Lieberman's program. Lurie spent a lot of time meeting with leaders of organizations that Tipping Point supported, listening to their challenges and frustrations, trying to understand how his organization could help.

In one such meeting I found myself talking to Lurie and Dr. Mark

Ghaly, the medical director at the county health clinic in Bayview. At one point Lurie asked us what we thought was the biggest problem in the community. The term *ACEs* was out of my mouth immediately, and Dr. Ghaly agreed that he was seeing the same patterns and connections between adversity and ill health at his clinic. Lurie asked what we would do about it if money weren't an object. Soon I was riffing, pie-in-the-sky-style, about a whole new center that would focus on working up new protocols and treatments for kids dealing with high ACEs and advocating for those solutions nationwide. Dr. Ghaly was enthusiastic and added some suggestions for how to make such a center the cornerstone of the community. At the end of the conversation, I could see the wheels in Lurie's head turning, which is always a really good sign.

A few weeks later Lurie called me to say that he had found a way for Tipping Point to help us raise the money to create a center. The organization was going to make our project the focus of the next year's benefit fundraiser. We would need to have a plan with a thoughtful budget and a clear vision of what we wanted to accomplish, but Tipping Point could help us get the money. It was time to put all our dreams down on paper. As Lurie talked, I was uncharacteristically quiet. This really was our chance, and this time, I would be fully prepared, not just with a statement of the problem, but with the solutions as well.

As soon as I got off the phone with Lurie, I called Victor Carrion. We talked through the kinds of resources it would take to pilot interventions for toxic stress. We dreamed of a sort of innovation lab that would do three things for our patients — prevent, screen, and heal the impacts of ACEs and toxic stress. The overarching goal was always to use the clinical science that came out of our center to change medical *practice*. To do that, we landed on a synergy between three pillars — clinical work, research, and advocacy. The clinical arm would be devoted to caring for patients and developing new approaches for treating toxic stress in a real-world setting. Research meant that we would hire a team to do what Dr. Clarke, Julia Hellman, and my other partners had been doing at the Bayview clinic — scouring the literature for best practices and using them to inform our clinical work. In addition, our research team would help us figure out how to validate the interventions and tools we were using and would always be on the lookout for ways to refine those practices according to the highest standards of medical science. Advocacy was the final piece. That was where we hoped to raise awareness and share the solutions that we'd found were working in our clinic so that eventually we might see broad-scale adoption by every pediatrician in America and beyond.

After putting some feelers out into the philanthropy world, we decided to join forces with Katie Albright, a tireless child advocate who was trying to create a center of her own that would offer complementary services. Housing both our organizations in the same building and fundraising as a unified front would be much more compelling to potential donors than each of us doing it individually.

Ebullient phone calls, illegible notes scrawled on the back of junk mail, and delicious spikes of adrenaline filled the days and weeks that followed as we fleshed out plans for what we would ultimately call the Center for Youth Wellness.

• • •

True to his word, Lurie had the Tipping Point throw its weight into funding our dream by making the benefit its biggest ever. The organizers hired a production company to make a jazzy video to promote the center's vision and somehow even managed to land John Legend as the benefit's headliner. The evening was a smash that I remember in surreal snatches of excitement and color. I wore a black vintage Oscar de la Renta dress scored from a consignment shop and my lucky four-inch heels that were hell on my musculoskeletal system but that made me feel like anything was possible. (When I got to sit next to John Legend at dinner, I made a mental note never to throw those shoes away.) Halfway through the night Lurie got up on the stage and introduced our plan for the center. The video completed his call to action, and he then started the bidding. The philanthropists of the Bay Area and the titans of tech responded, their glow sticks bobbing in the darkened room. The next thing I knew, Tipping Point had raised $4.3 million and John Legend had taken the stage and was belting out my favorite song. As a doctor I know you can't die of happiness, but when

I stepped out onto the dance floor in my lucky heels, for a moment it felt dangerously possible.

. . .

Now that we had the funding to start the project, we needed to figure out the steps to make the dream a reality. Dr. Carrion became a cofounder with me and it was a match made in heaven. We continued to think through approaches to treatments and research. Kamala Harris and Daniel Lurie loaned us experts from their teams to help us work out the details. Shortly after the benefit, we sat down and looked at the nuts and bolts of things, and we realized just how quickly the $4.3 million dollars would fly out the window when split among three organizations — the expansion of the Bayview clinic, the new Center for Youth Wellness, and Katie Albright's children's advocacy center. It had seemed like an enormous sum when I was celebrating on the dance floor, but with the crazy San Francisco real estate market, it wasn't even enough to buy a building. In fact, renting, designing, and renovating a 26,000-square-foot building, plus meeting the stringent federal codes for a health clinic, would eat up almost all the money.

As discouraging as it was to realize we weren't swimming in dough, we still had enough to get started. It was seed funding and it was enough to bring the Center for Youth Wellness (CYW) into the world. The Bayview clinic, supported in part by the hospital, would keep doing what it did — regular checkups for kids in the community and ACE screening. Once a patient screened positive for ACEs, the CYW clinical team would provide the multidisciplinary services focused on treating toxic stress — mental health, mindfulness, home visits, nutritional counseling, all the stuff our research told us could make a difference. The research team would track the data, and the advocacy team would get the word out. It was going to be a top-to-tail health-care home for kids and what we hoped would be a model for future organizations.

After a year of planning and fundraising for CYW, it was finally time to act on the business plan and build. In August of 2011, I transitioned from my role as medical director at the Bayview Child Health

Center to become CEO of the Center for Youth Wellness. The title *CEO* was aspirational at the time. There was not much for me to be CEO of — I was literally working out of my kitchen. I was fortunate to have the help of Rachel Cocalis, a recent college graduate and future lawyer who volunteered to work for free as my assistant until we became official and I could pay her. I was still seeing patients at the Bayview clinic but had scaled back to one day a week and had passed the medical-director baton to my colleague Dr. Monica Singer. My real job was focusing on the CYW plan and making it happen. The critical work of hiring a team meant doing interviews in coffee shops and at my dining-room table.

. . .

Although starting CYW was one of the scariest things I had ever done, it was actually going pretty well for a bare-bones operation. Which was why I was totally unprepared for what happened next.

Though we hadn't even opened our doors (in fact, we were still negotiating the lease on a building just a few blocks from the original Bayview clinic), we had to apply to the city for a change in the zoning code to allow the type of clinic that we were proposing. While it should have been a mundane process, funny things start to happen in Bayview when people hear that you have $4.3 million. Suddenly a small but determined group of individuals (six, to be exact) began to agitate and put up roadblocks. They didn't want us to locate our center at the site we'd found because they alleged that it was contaminated with "toxic dust." They had no evidence of contamination, but the rumor was enough to throw a huge wrench into the works. We paid for two rounds of environmental testing, which both came up clean. We even engaged the San Francisco Department of the Environment to do an independent sample that corroborated the findings of our experts — no toxic dust. But the group wouldn't be dissuaded. When the planning department granted our building permits, they appealed, triggering a three-month delay. I wanted to pull my hair out. We were under the gun to get the center up and serving children, but I felt I was wasting time and money jumping through hoops.

I would learn later that this is a common practice in low-income communities. When folks hear that there is money coming into the community, there is a small contingent that essentially makes its living by trying to get a piece of it. That the community would benefit by having more high-quality services for kids is not what they were interested in. They wanted the money *in their pockets*. These folks create problems for the team running the project, often using race as a lightning rod, and then they're conveniently available to be "community consultants" who can help the project move forward for a hefty fee.

While I understood the impulse to "get yours" when there was not much to go around, we weren't some multimillion-dollar corporation with money to burn. This group of six was focusing on a number that was misleading. Yes, Tipping Point had raised $4.3 million for the *entire project,* but what was easily missed, unless you were in our meetings crunching numbers, was that these dollars were being split three ways. After paying for the rent and construction, we had almost nothing left, and we still had to pay the staff. Clearly, this group had an entirely wrong idea about how deep our pockets were.

One afternoon, a staff member walked into the temporary offices that CYW had rented next door to the Bayview clinic. In her hand was a flyer that said STOP THE MASSACRE! DR. BURKE WANTS TO EXPERIMENT ON OUR CHILDREN!

I went quiet for a moment, taking stock of what was happening in front of me. My mind cycled through a few choice expletives that I had to try hard not to vocalize. Accusations of medical experimentation in African American communities are exceptionally loaded because they are founded on a history of shameful and unethical exploitation of blacks by the medical community. As this group undoubtedly knew, calling up that history preyed on people's legitimate fears, triggering a long-held mistrust of medical professionals. It burned me up that they would use the trauma that came out of those situations for their own purposes.

I quickly went online to check out the community message boards and saw posts and articles about why people in the neighborhood shouldn't trust "that Jamaican." If I hadn't been so upset, I almost would have laughed at the genius of it. Instead of playing the race card, they went the foreigner route, casting me as the malevolent outsider. I thought of my patients or their parents reading those signs and felt my chest tighten up and my face go hot. It took a minute for me to calm down, but I tried to convince myself that anyone in Bayview who knew me would know that this was total crap.

Until then, I had been trying to placate this group by jumping through all the hoops they set up. Now I realized it was time for a different approach.

I would need to meet one-on-one with the leader of the group, an eighty-four-year-old chain-smoking force of nature I'll simply call Sister J. I had heard stories about her from my patients' parents and others in the community for years, but until that moment I'd never been on the receiving end of her "advocacy." Sister J had lived in Bayview much of her life and was a legend in her own right. A longtime activist, she had done quite a lot of good for the community. She had battled environmental issues and advocated for fair housing and jobs. Unfortunately, with her, the line between community benefit and personal benefit could get a little murky. When the City of San Francisco was moving forward with the largest municipal solar-power system in the nation, she threatened to hold up the project and insisted that Bayview residents do the work. While she did win a good number of jobs for residents of Bayview, one of the concessions included a free solar-power system for her house. At other times, the benefit to the community was less clear. When San Francisco tried to implement gun-safety measures aimed at decreasing the number of kids who were victims of gun violence, Sister J was the lead plaintiff in the NRA-supported legal effort to stop the legislation. She claimed her Second Amendment rights were being violated.

There were those on my team who wondered aloud if we should just give in to the game and "hire" her as a consultant. My answer was simple: Over. My. Dead. Body. I wasn't about to use our limited dollars to buy into a vicious cycle of exploitation. My goal in meeting with her was to explain what we were trying to do and why it was so important. I knew that at her core she cared about the community, and I hoped that if she understood that we didn't actually have a ton of money and were just trying to bring services to help kids, maybe she would cut us some slack.

It wasn't long before I found myself nervously ringing Sister J's doorbell, trying to get the stop-the-massacre flyer out of my head. I wanted to exude a sense of calm and solidarity. No small feat. When she opened the door, I had to look down. She might have been a big presence, but Sister J was just about five feet tall, with deep creases in her soft face and glasses that hung on the end of her nose. She looked the part of a matriarch, like one of those Southern grandmothers who knew how to keep generations of family together and made sure that everyone knew "our history." She politely invited me in and we sat down in a perfectly appointed living room on a settee preserved for all time under a thick plastic cover.

Before I could say anything she handed me a business card that gave her title as Community Icon. I looked up and searched her face for concealed amusement, evidence of what I could only imagine was a self-deprecating joke. Instead, she poured us both tea and began to talk.

The power dynamic was palpable. The tea and manners were her subtle way of letting me know who was boss. Her voice was gravelly from decades of smoking, but she proceeded to hold forth for the next two hours.

Almost uninterrupted, she told me the story of her life. I understood that this monologue was meant to communicate her bona fides — what she had done for the community and why she was so respected (and feared). But I was distracted by a profound irony — her life was riddled with ACEs. The mental tally I had going in my head put her ACE score at a seven or eight by the time she wrapped up.

Finally, I had a chance to tell her why I was there. I started to explain everything that I'd seen in my patients, why the work was so important to me, and how much I thought that we could lift up not only Bayview, but many communities around the country and the world that were profoundly affected by ACEs. Before I got very far she interrupted and started talking over me. It was clear that I was there to listen, not to talk. This was never intended to be a two-way conversation. I took a deep breath and considered my options. It wasn't looking good in terms of changing her mind about our building, and part of me wanted to down my tea and leave, but I decided to dig in and keep trying. She was the person standing between my kids and the dream of the Center for Youth Wellness. I let her continue for a few more minutes, and she got to her final tale of activism.

"I told them that I was going to blow that building up . . . but I wouldn't do that to you, baby," she said, finishing with a chuckle.

Out of nowhere, tears welled up in my eyes and spilled down my cheeks.

It wasn't the veiled threat or the disappointing lack of dialogue that upset me; it was the utter futility of the past several months that I had spent trying to work with this group. I believe in the power of conversation, connection, and empathy when it comes to dealing with community problems, but I had finally hit a situation where that just flatout didn't work. I could have been Nelson Mandela and it wouldn't have mattered to Sister J; her own agenda was the only one she was interested in.

She started to talk again, but for the first time, I interrupted her.

"I think we can do better by our kids," I said, rising to my feet.

I could see her eyes narrowing, but before she could say anything I continued. "Sister J, our kids deserve better."

And with that, I shook her hand and walked out.

• • •

For the next couple of nights, I couldn't sleep. I had a copy of the Stop the Massacre! flyer on my nightstand next to my bed, and every night as I lay down, I felt my heart start to race. How many people had seen that flyer? There were so many people in the neighborhood I hadn't yet met. Did anyone really believe that I was experimenting on kids? Rumors are like termites in small communities like Bayview; they work fast and do a lot of damage. Worse yet, how would the planning commission react to the allegations? I had no idea. I was beginning to see that the lack of outside investment in Bayview wasn't only because nonresidents didn't care; even the people who did care had to deal with ridiculous obstacles placed in front of them by a few misguided gate-keepers. I could see how easy it would be for anyone trying to do some good in Bayview to give up.

Fortunately, a few nights before the planning commission hearing, I got a call from the author and journalist Paul Tough. Before the whirl-wind run-up to the launch of CYW, he had written an article in *The New Yorker* about the Bayview clinic and our work around ACEs and toxic stress. Being more of a medical-journal gal, I had no idea what a big deal this was until the issue hit the stands. It's not overstating the situation to say that the article changed everything. By spotlighting the subject, it triggered a ton of interest among colleagues and new sup-porters and brought our work into the mainstream. Paul and I had de-veloped a friendship over the weeks and months he'd spent walking with me to work and shadowing me at the clinic, so occasionally, he would reach out to see how things were going, and this time I spilled about Sister J. A few minutes into my sob story, I took a breath and heard a knowing laugh on the other end of the line.

"*What* could you possibly be laughing at?"

Paul told me that Geoff Canada, founder of the Harlem Children's Zone and one of my personal heroes, had also faced some pushback from community members when his organization was building a new school and community center in the middle of a housing project in Harlem. Tough had written the book on the legendary educator and his organization's work to transform educational outcomes for chil-dren in Harlem. Canada found that the opposition evaporated when people saw that the Harlem Children's Zone was there for them and that the building and the organization was an asset.

"It's a rite of passage," Paul assured me. "You'll make it through. Consider it a badge of honor."

. . .

After my conversation with Paul I was able to step back and get a little perspective. It occurred to me that the trauma that is endemic in com-munities like Bayview isn't just handed down from parent to child and encoded in the epigenome; it is passed from person to person, becoming embedded in the DNA of the society. That was exactly the kind of cycle we were hoping to break with our work at CYW. That realization caused me to look at this obstacle as a symptom of a community plagued by trauma as opposed to a sign that I was destined to fail. Paul also reminded me to stick to what I already knew: my patients and their parents were overwhelmingly supportive of our plan for CYW. Happy parents were constantly referring relatives and friends to us and asking when we were going to hire more doctors and therapists. They had seen firsthand the good we were doing in the community. We knew that a small but strident group of folks would oppose our application during the planning commission meeting, but I also knew that a hell of a lot more folks wanted to see us open the doors to a bigger and better facility. I needed to focus on harnessing that strength instead of worrying about the opposition.

In the days that followed my conversation with Paul, my team and I started to talk to our kids' parents and others in the community. We let them know that the project was in jeopardy and that we needed them to show up at city hall. On the day of the hearing, people set up carpools and we got vans to help bring in our supporters who didn't have transportation. Many folks had to take the day off work. For their time and effort, the best we could do was provide lunch — Subway sandwiches. As people arrived, we gave them green stickers to wear to signify their support of our project. When the meeting started, the room was packed and the crowd spilled out into the hallway. The members of the planning commission made their way through the agenda and finally got to us. A small handful of people got up and spoke out against the project.

Then it was our turn.

Family after family rose and testified. They were every shape and size and all shades of the rainbow. Some had brought their kids, and all of them talked about what we had done for their families, what it meant to them, and how much additional services were needed. With each person who spoke, I felt my body relax and my chest open up. At one point I looked over at my team and just shook my head. It was one thing to hear that kind of gratitude in the privacy of the clinic; it was another to hear it proclaimed publicly and with such feeling. In that moment, my faith in our work deepened. Here in front of me was the blueprint for our success—a community of people struggling with a legacy of ACEs, up against obstacles strengthened by historical cycles of marginalization and violence, but nonetheless coming together to advocate for a better life for their kids. These families testified that we were doing something powerful and important for their children. The cycle could be broken. Kids were staying in school instead of hitting the streets. Parents were learning to talk to their kids instead of disconnecting from them. This group in front of me saw an opportunity in CYW for their families and their community to further the process of collective healing. I realized that CYW already had the most important ingredient for success: the trust and support of the community we aimed to serve.

. . .

After everyone had spoken, the planning commission asked those who were opposed to CYW to stand up.

Four lonely figures got to their feet.

Next, the commission asked everyone who had come in support to stand up.

A sea of green stickers rose in unison, an extended family of over two hundred supporters—patients, parents, staff, friends, and family. Overwhelmed, I was struck yet again by the way that people in our community take care of one another. That moment is what Bayview looks and feels like from the inside, and I have to say, it feels pretty damn good.

When the planning commission voted unanimously in our favor, a wave of raucous cheers swept the room.

9

Sexiest Man Alive

FOR MOST PEOPLE, the name Dr. Robert Guthrie doesn't set their hearts aflutter, but as my brothers like to remind me, I might be a special case. In my mind, Dr. Guthrie is right up there with JFK Jr. and Idris Elba. Definitely on my short list for the "name any person, dead or alive, you most want to have dinner with" game. I don't know if *People* magazine was around in 1961, but if it was, the development of newborn screening should have earned Dr. Guthrie a spot on the cover of the "Sexiest Man Alive" issue.

I first heard of him when I was a young medical student learning about newborn screening, which is an important way for doctors to identify a long list of life-threatening diseases like hypothyroidism and sickle cell anemia. Anyone who has had a baby may remember that at some point, after about twenty-four hours, the baby's heel gets poked and a drop of blood is collected so the lab can do what's called a newborn screen. This test allows doctors to identify disorders (like hypothyroidism) long before symptoms develop and then treat the underlying issue before it can cause problems. This leads to much better outcomes for patients, and it's now the standard of care in every developed country around the globe. But that wasn't always the case.

. . .

Dr. Guthrie started his career as a cancer researcher, but his life changed in 1947 when he and his wife, Margaret, had their second child, a son they named John. Not long after John was born it became clear that he had a significant mental disability, what was referred to at the time as "mental retardation." Despite taking him to specialist after specialist, the Guthries never learned the cause of John's disability. After his son was born, Guthrie dedicated himself to the prevention of mental disabilities. By 1957, he had become the vice president of the Buffalo chapter of the New York State Association for Retarded Children. The following year, Margaret Guthrie's sister, Mary Lou Doll, had a baby girl whom she named Margaret after her beloved sis. At first, Margaret was the picture of a happy infant, doing all of the smiling and cooing that the books tell you to expect. But over time, baby Margaret's demeanor changed. She became quieter and less interactive, and by seven months, she began losing her milestones and developed an odd habit of dropping her head. Concerned, Mary Lou Doll took her daughter to her pediatrician, who diagnosed Margaret's "head dropping" as seizures and determined her to be "somewhat retarded." Although there was a test available at that time for the rare genetic disease phenylketonuria (PKU), it wasn't done. The pediatrician did, however, recommend a brain wave test, although he said there was no hurry "as she was very young for specific results."

It wasn't until Margaret was a year old that Mary Lou conferred with her brother-in-law about her daughter. He suggested she take the baby to the University of Minnesota, where she was finally tested and diagnosed with PKU. Phenylketonuria is caused by an enzyme deficiency that renders the body unable to metabolize phenylalanine, an amino acid found in most proteins, including breast milk and baby formula. Over time, a byproduct of phenylalanine builds up in the body and slowly poisons the developing brain and nervous system. Margaret Doll's seizures were a result of a toxic buildup of this phenylalanine byproduct. There is a treatment for PKU, but here's the kicker — it isn't a million-dollar-a-dose medicine or some fancy implantable medical device. To prevent the neurotoxicity of PKU in a child, all you have to do is stop feeding the kid anything with phenylalanine in it. If you've ever read the fine print on a can of diet soda, you might have wondered why it says "This product contains phenylalanine." That piece of information is meant to help people with PKU maintain the phenylalanine-free diet that is so critical to their health.

Margaret Doll was started on a phenylalanine-free diet when she was thirteen months old, and over time she regained some of her developmental milestones. She sat up at eighteen months and began walking when she was two and a half years old, but she remained severely intellectually disabled, with psychologists reporting her IQ as twenty-five.

The combined heartbreak from his son and his niece made Dr. Robert Guthrie a man on a mission. He knew that if the PKU was caught early enough, the phenylalanine-restricted diet would prevent severe neurological damage. At the time, PKU was diagnosed by what was called the diaper test, which involved testing for the toxic phenylalanine byproducts in the urine. Though the test was accurate, it was not sensitive enough to detect the toxic byproduct until after severe brain damage had already occurred.

Guthrie took it upon himself to find a better method of measuring blood phenylalanine. Borrowing methods from his experience in cancer research, he was able to devise a test that required only a few drops of blood. The blood was placed on a piece of filter paper, then the filter paper was put in a culture of bacteria that would grow only in the presence of phenylalanine. If bacteria grew, he knew there was phenylalanine where there shouldn't be.

In 1960, one of the first trials of the Guthrie test was done on children at the Newark State School for the Mentally Retarded. The test confirmed every single known case of PKU as well as four that had gone undetected. Soon after, Guthrie set up a laboratory near the Buffalo Children's Hospital, and over two years he went on to test more than four hundred thousand infants from twenty-nine states for PKU. The new screening method identified thirty-nine cases of PKU in newborns, and treatment was started early enough to prevent brain damage. Further, the test didn't miss a single case of PKU.

• • •

For years after he developed the test, Guthrie was a vocal advocate of screening all newborns for PKU before they left the hospital. He fought side by side with like-minded organizations to demand that the test be mandated by law. He succeeded, and eventually the newborn screening test was expanded even further; it now identifies more than twenty-nine conditions that can lead to long-term neurologic damage. The Guthrie test has been used in more than seventy countries and is responsible for helping countless children reach their God-given potential. Now, if that doesn't earn you the "Sexiest Man Alive" title, I honestly can't imagine what does.

To me, Guthrie's true legacy was that he set the precedent for universal screening. It's something I think about every time I see an ACE score in a patient's chart. In the same way that babies with PKU aren't born with any outward signs that they have the genetic disorder, kids don't come into my office with signs around their necks saying I HAVE TOXIC STRESS. That's why the *universal* is just as important as the *screening*. Time and again, I am reminded of what Guthrie showed the world — that we shouldn't wait for our kids to come to us with symptoms of neurologic damage when there's something simple that can be done to prevent it.

. . .

Three years after we opened the Center for Youth Wellness, I began seeing a new patient who brought Guthrie's lesson home yet again. Lila was two and a half years old, blond, bubbly, and precocious. One day in the fall of 2015 I sat at the conference table with my colleagues sipping tea and reviewing her chart. Once a week CYW has multidisciplinary rounds; that's where we discuss treatment plans for patients who have been identified by the clinic as being at high risk for toxic stress. This approach to care was something that started in the Bayview clinic out of necessity.

In the early days of the Bayview clinic, I was overwhelmed not by the workload (although that was some craziness too), but by the dire situations my patients and their families were often in. I was trained to treat asthma and infections, but my patients needed so much more than prescriptions for inhalers and antibiotics. Sometimes they needed housing, protection from abusive parents, or even things as simple as basic toiletries. One day I had a patient's dad tell me that his family had been thoroughly burglarized; the person who broke in had even taken the toilet paper off the roll. (You know you have been good and robbed when someone takes your freaking toilet paper.) That same dad proceeded to board up the windows to prevent another break-in. Soon after, all three of his children came to me on the same day with severe asthma exacerbations, and the dad asked, sincerely, "Hey, Doc, do you think it's bad for their lungs that we're smoking meth in the house with all the windows boarded up?"

That same week, a seven-year-old patient was brought to me with a complaint of chronic headaches. She had just been removed from her uncle's home, a studio apartment where she had literally watched her uncle sexually abuse her fifteen-year-old cousin, his daughter.

Back then I dictated my notes into a tape recorder, and when I listen to them now, I swear my heart remembers, aching yet again with grief for my tiny patients. There were days I would walk out of the exam room, close my office door, lay my head down on my desk, and just cry. And I definitely wasn't the only one. At lunch or after work, I would find myself talking about my patients to Dr. Clarke and our social worker Cynthia Williams, partially to blow off steam but also because talking to one another helped. We would put our heads together to find avenues of support for our patients, which was good for them and for us.

Eventually, I realized that what we were doing at the clinic was an informal version of a practice I had learned on the oncology ward at Stanford, referred to as multidisciplinary rounds. In the pediatric oncology unit, there are understandably some really high-needs patients. Every week a group would meet that included the head oncologist, the social worker, the therapist, the child-life specialist (someone who helps kids through painful procedures), and a nephrologist (kidney doctor) or whatever specialists were needed for that particular case.

It was a perfect example of divide and conquer. When you're caring for kids with cancer, by definition you have an incredibly sensitive and complex situation — *of course* no one person (doctor or otherwise) can adequately address all of those needs. When I thought about our patients at the Bayview clinic, their needs didn't seem too different in terms of complexity of care. So instead of bellyaching in the break room, Cynthia Williams, Dr. Clarke, and I began meeting every week, bringing a stack of charts to review and calling it, Stanford-style, MDR.

Straight out of the gate, we could all feel that the practice made a huge difference. It allowed me to do my job well without having to split my energy or wear multiple hats. I knew when I walked into an exam room that I would have a place to bring all of the challenging issues at home that were also affecting my patients' health. I didn't need to be a social worker or a therapist; I could let Williams and Dr. Clarke do their jobs in a way that coordinated with what I was doing in the exam room. As a result, my patients got a better doctor, and their additional needs were addressed by someone who was trained to take care of them.

We weren't aware of this at the time, but our approach would later become a best practice known as team-based care. Our patients' lives didn't get less complicated, but we found that this new model helped patients get better faster, and it had the added bonus of improving staff morale (especially mine). It was such a success that when we opened CYW, it was an important priority to carry that practice forward.

Years later, as I looked around the table at CYW, I felt a sense of pride and confidence seeing two social workers, a psychiatrist, a clinical psychologist, a nurse practitioner, and two wellness coordinators whose job was to manage the interlocking web of patient treatment plans across disciplines. I was about to give all of them the scoop on what turned out to be my most unexpected patient in months, and I knew that together we could help her.

· · ·

When Lila first came into my exam room, she was just tagging along with her baby brother, Jack, who was there for a follow-up appointment after a trip to the ED for an ear infection and a bad cold. It was the third ear infection for the nine-month-old and he had also had two bouts of pneumonia. His parents wanted to make sure that this cold wouldn't "turn into" another pneumonia. Lila was the same age as my son Kingston (yep, managed to find a husband and have a baby in the middle of all this). I laughed as she scampered around the exam room and asked precocious questions, just like Kingston did every morning as I was getting him dressed.

The family was new to the area, having just moved to the Bay Area from Ohio, so once I checked the ear (it was fine) and listened to his lungs (all clear), we scheduled physicals for both Lila and her brother. The family was freaking adorable. I see a lot of beautiful families in my line of work, but these guys really stood out. Molly and Ryan were young parents, but completely doting and caring, and to me they seemed as close-knit as the Cleavers. In the course of my exam, I encountered a soiled diaper (occupational hazard), and Ryan jumped up to change it, apologizing profusely. It was sweet to see both parents so hands-on with the kids.

Two weeks later when I saw that both kids were on my schedule for their physicals, I smiled instantly. I was looking forward to seeing them again. I walked into the room and reviewed all of the standard intake and medical-history forms. I was glad to see that Jack had no new symptoms and noted that Molly's only concern was Lila's growth. Ryan wasn't able to make the visit, so it was up to Molly to explain her daughter's history. She told me that when Lila was born, she was at the twenty-fifth percentile for height and weight, but over the subsequent six or so months, she had drifted down below the third percentile and stayed there. Their previous pediatrician had counseled them about diet and even recommended PediaSure (a nutritional supplement), but nothing seemed to work. Molly didn't understand why Lila was so small. Both she and Ryan were average height and Lila had never had any chronic health problems. As I sat down with her to finish reviewing the medical-history forms, I flipped to the ACE score and had to force myself not to do a double take.

. . .

Maybe Mom misunderstood the directions? I thought. *Maybe she wrote down her own ACE score instead of the kids'.* According to the paperwork, Lila had an ACE score of seven and her nine-month-old brother's score was five.

I started my usual spiel, figuring Molly would realize her mistake: "New research has shown that children's exposure to stressful or traumatic events can lead to increased risk of health and developmental problems, like asthma and learning difficulties. As a result, at this clinic we now screen all of our patients for adverse childhood experiences. I'm going to review this list of ten items, and you don't have to tell us which ones your child experienced, only how many. I'd like to take a moment to go over your responses." Molly was nodding her head the whole time.

"I totally believe it," she said.

"So you've heard of ACEs before?" I asked, a little perplexed.

"No, but when I read about it on the piece of paper, it made total sense."

She confirmed that her kids' ACE scores really were seven and five. *So she hadn't filled out the form wrong.*

The realization knocked me in the gut. I see patients with high ACE scores, even some that are pretty young, every day, and it's always tough. But Lila's mannerisms reminded me so much of my own son's that her ACE score hit me in a way that I hadn't anticipated. The doctor in me was grateful to get the insight into what might be going on with her health, but as a mother, it made my stomach sink. I wanted to throw my arms around Lila, hold her tight to my chest, and tell her it would be okay. I wanted to make those seven ACEs disappear like kissing away one of Kingston's boo-boos. But I couldn't. And it wasn't my role. What I could do was make sure that Lila's ACEs weren't written into her biology for the rest of her life. In fact, that was my job.

I knew from Lila's ACE score that she was at much greater risk for a host of adult health problems than other kids. But what did this information mean for how I should do my day-to-day job? Felitti and Anda had looked at health outcomes in adults, but Lila was unlikely to face most of those diseases for decades to come. Fortunately, our research team at CYW had made good progress in filling in some of those blanks.

Our team reviewed over sixteen thousand research articles on the impact of childhood adversity on health. What we found was that childhood adversity is associated with a variety of diseases and conditions in children that can be observed as early as infancy. In babies, exposure to ACEs is associated with growth delay, cognitive delay, and sleep disruption. School-age children show higher rates of asthma and poorer response to asthma rescue medication (such as albuterol), greater rates of infection (such as viral infections, ear infections, and pneumonia), and more learning difficulties and behavioral problems, and adolescents exhibit higher rates of obesity, bullying, violence, smoking, teen pregnancy, teen paternity, and other risky behaviors such as early sexual activity.

I sat down to walk Molly through what I suspected was going on with her daughter's health.

"I think that because of what Lila has experienced, her body may be making more stress hormones than it should, and this may be affecting her growth," I said.

This seemed to make intuitive sense to Molly.

"Yeah. We were working on her weight with her previous pediatrician. Her dad had times when he would be away from the house, and it seemed like when he was away, her weight would pick up a little bit, but when he came back, it would fall off again. There has definitely been a lot of stress in our house."

"Wow. Did you ever mention this to her previous doctor?"

"No," she replied. "He never asked."

If it wasn't for the ACE scores, no one would have suspected that Lila and her brother were at such high risk for so many health and developmental problems. Possibly they might have gotten some attention if they'd started to show behavioral problems in preschool, but even in that case, it's likely they would have been diagnosed with ADHD and been funneled down the medication path. If they had never manifested any behavioral symptoms, chances are — even if they developed asthma, or an autoimmune disease, or any of the other significant immunological consequences of toxic stress — the *underlying* problem would likely have gone undetected and untreated. Guthrie had shown that the only way to radically move the needle on patient outcomes is to screen *universally,* because otherwise you are relying on chance: The chance that Lila's symptoms would get bad enough that her doctor asked more questions. The chance that this particular doctor had heard about ACEs and knew to ask those questions in the first place. How much damage could be done while you were waiting for the right questions to be asked, the right tests to be run? Guthrie knew. His sister-in-law knew. They saw what happened when PKU was not tested across the board, when the opportunity for early intervention was lost. That is why an ounce of screening is better than a pound of cure.

. . .

In the case of PKU, it's clear that early intervention is needed to treat the condition successfully, but what about ACEs and toxic stress? It's actually just as clear. All the science about the development of the neuro-endocrine-immune system tells us one thing: intervening earlier is better (and I mean way, way, *way* better). That's not to say that older kids and adults with ACEs can't benefit from interventions (more on that to come), but the later we start, the more intensive (and expensive) the treatment has to be and the less likely it is to be effective. The reason for this is that starting earlier gives us more tools to work with.

The past several decades of neuroscience research explains why early adversity has such an outsize impact on children's development. The prenatal and early childhood periods offer special windows of opportunity because they represent "critical and sensitive periods" of development. A *critical period* is a time in development when the presence or absence of an experience results in irreversible changes. Much of what we know about critical periods comes from research on binocular vision (the ability to perceive depth and create a 3-D image out of inputs from both eyes). When a baby is born with eyes that are misaligned (with crossed eyes or a lazy eye), the brain will have trouble creating a coherent 3-D image and depth perception is impaired. But if the misalignment is identified and corrected by age seven or eight, the child can go on to develop normal binocular vision. After age eight, however, the window closes and the opportunity for normal 3-D vision is permanently lost. (Or so we thought — new data suggests that the window for binocular vision may be longer than previously believed, and exciting research is focused on learning if we can re-open windows previously thought to be closed.) Since the discovery of critical periods in the brain's visual cortex, scientists have found that numerous other brain circuits demonstrate critical periods as well.

A *sensitive period* is a time when the brain is particularly responsive to a stimulus in the environment, but unlike critical periods, the window doesn't totally close at the end of the sensitive period; it just gets a lot smaller. The development of language is a great example of a neural circuit that demonstrates a sensitive period. Everyone knows that it's way easier to learn new languages when you're a kid than when you're an adult. I have some European friends whose kids speak four languages fluently — English, French, German, and Spanish — each with a flawless accent. Meanwhile, several years and hundreds of dollars on Rosetta Stone later, my French is *très terrible.*

Critical and sensitive periods are times of maximal neuroplasticity (the brain's ability to rewire or reorganize itself in response to a stimulus). This growing and changing of neurons and synapses can happen in response to injury, exercise, hormones, emotion, learning, and even thinking. Our brains are always changing in response to our experiences, and overall, that's a good thing.

There are two types of neuroplasticity, cellular and synaptic. *Synaptic plasticity* is a change in the *strength* of the connection across the junction from one brain cell to the next (the synapse). It's kind of like changing your voice from a whisper to a shout. *Cellular plasticity*, however, is a change in the *number* of brain cells that are talking to each other, the difference between one person shouting and a whole stadium shouting. While synaptic plasticity is lifelong (it's how an old dog learns new tricks), cellular plasticity happens most rapidly in the first years of life. About 90 percent occurs by the time a child turns six, but the rest of it stretches out until about age twenty-five.

The way brain development works is like those weird topiary bushes that grow in the shape of Mickey Mouse or a giant dinosaur (stay with me here for a second). Obviously, they don't just grow that way on their own; they are pruned. Babies are born with an oversupply of brain cells and the brain also goes through a pruning process. The brain cells on the circuits you *don't* use get pruned, and the ones on the circuits that you *do* use grow and strengthen. Our experiences, both positive and harmful, determine which brain pathways are activated and continue to strengthen over time. In that sense, early experiences literally shape the brain.

We know that early adversity activates the brain pathways that are associated with vigilance, poor impulse control, increased fear, and inhibition of executive functioning. But if we can identify kids who are at high risk for toxic stress early enough, we can intervene in time to take advantage of high levels of both synaptic and cellular plasticity. The most effective way to rewire the brain is to implement early interventions that help to prevent the stress response from becoming dysregulated and that support practices that buffer the stress response (as with child-parent psychotherapy). By doing this, you give the brain the greatest opportunity to grow in new and healthy ways.

So what about all of us old dogs? Well, when it comes to learning new tricks, the good news is that the hormonal changes occurring in adolescence, pregnancy, and new parenthood open up windows of neuroplasticity that are believed to be additional sensitive periods. Testosterone (in boys) and estrogen and progesterone (in girls) are sex hormones that lead to all of the mortification associated with adolescence (acne, body hair, breasts, menstrual cycles). Another important hormone is oxytocin, a powerful bonding hormone that is released in very high levels by the mother during childbirth and in the immediate postpartum period. All of these hormones stimulate synaptic plasticity, biochemically enhancing the ability to learn and adapt to one's environment. These times represent special opportunities for healing, moments when enriching experiences have an even better shot at being "wired in."

More good news — there are things that you can actually do yourself to boost your synaptic plasticity; sleep, exercise, nutrition, and meditation all enhance the process. That being said, a little more patience and consistent practice is required for adults, since the change will not be as radical or as fast as it is in young children. We know that the earlier we start, the more tools we have — young children are the most vulnerable to adversity, but they also have the greatest capacity for healing when the interventions are begun early. And we also know that it's never too late to use biology to our advantage for healing.

. . .

Guthrie famously developed the simplified blood phenylalanine test in three days. Unfortunately for us at the clinic, developing a fast and easy screening protocol for ACEs was anything but fast and easy. By 2015, we'd been working on it in one way or another since 2008. At the Bayview clinic we started by simply asking about patients' history concerning the ten ACEs and recording that information in their medical charts. The problem with this approach was that it took a long time and sometimes meant that the doctor asking the questions had to navigate a serious emotional obstacle course that most primary-care clinicians have neither the time nor the training to navigate thoughtfully. Though it helped us provide better care for our patients, it wasn't ideal. We knew we had to make some adjustments if it was going to work for doctors outside our own little clinic.

The great thing about CYW is that it is built on the successes of the Bayview clinic. We were on the right track in terms of screening, and once CYW had the resources, our clinical and research teams put their heads together to refine the screening tool so it could work for every doctor. It needed to be simple to use and evidence-based.

Fast-forward a few years (not to mention some sweat and tears, but fortunately no blood). The screening tool that Lila's mom filled out was much different than the one I'd first used with my patients. First, the new one was on paper (or a tablet), something that a parent was able to fill out before I came into the exam room. Second (and this was the real innovation), on the new one, we listed the ten ACEs and specifically asked the patient's parents *not* to tell us which of them their child had experienced, only how many. At the bottom of the page, the caregiver wrote the total number, and that's the ACE score. We call this our "de-identified" screen because it doesn't identify the individual ACEs, and it goes a long way to solving two of the biggest challenges — time (previously, a positive screen took a very long time to unpack) and the sensitive information we're asking for. As both Dr. Felitti and I had seen firsthand, doctors, more than patients, are hesitant to get into conversations about past incidences of abuse or neglect. They worry their patients will be uncomfortable, that they won't tell them the truth, or, worse, that they will tell them the truth and the visit will be derailed with an emotional outpouring or the need to file a report with Child Protective Services. The de-identified screening tool takes all of these concerns out of the equation.

The other important thing that the CYW ACE questionnaire did was go beyond the traditional criteria developed by Felitti and Anda and ask about additional risk factors for toxic stress. We don't call them ACEs because they are not from the ACE Study and we don't have the large body of population data to tell us odds of disease, but our experience in Bayview told us that our patients faced other adversities that repeatedly activated their stress-response systems. Our research team worked actively with the community (youth and adults) to learn what the greatest stressors were in their day-to-day lives. Informed by these insights, we reworked our screening tool to include other factors that we believe may also increase the risk for toxic stress.

- Community violence
- Homelessness
- Discrimination
- Foster care
- Bullying
- Repeated medical procedures or life-threatening illness
- Death of caregiver
- Loss of caregiver due to deportation or migration

In our teen screener, we also include the following:

- Verbal or physical violence from a romantic partner
- Youth incarceration

We score these supplemental categories separately so that we don't lose the ability to apply findings from the scientific literature. I know from the ACE Study that if a patient has an ACE score of four or more using Felitti and Anda's criteria, he is twice as likely to develop heart disease and four and a half times as likely to become depressed. Researchers are just beginning to look at the supplemental categories on a large scale, but the preliminary data indicates that stressors at the household level (the traditional ACEs) seem to have a greater effect on health than stressors at the community level. This was a surprise to many in the field (myself included), but the data suggests that if a child grows up in a stressful community environment but has a well-supported and healthy caregiver, he or she is much more likely to stay in the tolerable stress zone as opposed to the toxic stress zone.

When I reviewed Lila's screen, all I saw was that her score was a seven plus zero (seven for the traditional ACE score and zero for our supplemental score). That was enough information to tell me what I needed to do next. Molly didn't have to disclose any of the details of what had happened in their family if she didn't want to. And for the most part, she didn't. She mentioned only that Ryan had spent some time in rehab and that he had a history of ACEs himself. As I looked at Lila's ACE score, part of me wanted to know the whole backstory. I wanted to know how this dad who had been so happy to jump on a dirty diaper might be harmful. I wanted to know about this mom and what her story was. But to do my job well, I couldn't be the one to unpack that. In order to make sure that my twelve other kids on the docket for the afternoon were also screened for ACEs, I had to trust my team to take it from here. The de-identified screen allowed me to recognize that Lila's failure to thrive was most likely due to toxic stress. I needed only to get her the right care in a way that was fast and easy enough that I could reliably do it for every single one of my kids without being in the clinic until midnight every night.

. . .

I brought Lila's case to multidisciplinary rounds with a recommendation that she start child-parent psychotherapy. Ultimately, Molly would be best served by getting into the nitty-gritty of her daughter's ACE score with Dr. Adam Moss, Alicia Lieberman's most recent postdoctoral fellow. Her treatment involved three simple steps. The first and most critical was simply helping Molly better understand the problem and what we could do about it — going a little deeper into how stress hormones affected growth and how Molly herself had the innate capacity to be a buffer for her daughter's stress response. To make that happen, we had to help Molly learn how to get her own stress response in good working order. We explained to her later that we had a specialist who would teach her how to be a healthy buffer for her child's stress. The second step was getting mother and daughter plugged in to CPP, and the third step was just good old-fashioned PediaSure, which I thought would be more effective once we had dealt with the underlying toxic stress. Within three months, Lila was back on the growth curve.

When I think back to the early days of Diego and how overwhelmed I felt before we began the team-based approach, I knew this was a better way for everyone. Now, seven years later, it felt like this was how it had always been. It just made so much sense.

Unfortunately, sometimes what makes sense doesn't always align with the reality of medical practice. Another badass medical "she-ro" of mine is Sue Sheridan. While she is not a medical doctor, like Guthrie she has a son with a severe disability that inspired her to work tirelessly on behalf of families like hers. Most people have either had or know of a baby who was born with jaundice, a condition in which the infant's skin and eyes appear yellow. You might have even seen a picture or two of a friend's baby under the phototherapy lights, looking like he is in the newborn equivalent of a tanning bed.

Over 60 percent of newborns develop some amount of jaundice. The signature yellow skin lets pediatricians know that there is a buildup of a chemical called bilirubin in the baby's system. Bilirubin is created when the body breaks down old red blood cells. It is naturally processed by the liver and excreted by the body (which is actually why your pee is yellow). But when babies are born, it takes a little while for their livers to come online and function to full capacity, so the bilirubin can build up. Bilirubin is typically harmless, but if the levels get too high, it can cross the blood-brain barrier and cause brain damage.

When Sue Sheridan's son Cal was first born, he appeared as healthy and beautiful as a baby could be. But within the first twenty-four hours of Cal's life, his skin started to turn yellow. Sue and her husband were told not to worry, as jaundice was quite common in infants. No bilirubin test was done. At the time, standard of care was to do a visual inspection, meaning the pediatrician would eyeball the patient and decide if the jaundice looked severe enough to treat. Even though a blood test existed to measure bilirubin levels, it wasn't used routinely. The following day, Cal's yellow color continued to deepen, and though the Sheridans again expressed concern, still, no test was done. When Cal was discharged at thirty-six hours of life, he was described as having jaundice from head to toe, but his parents were simply given a pamphlet as they left the hospital that suggested putting the baby near a window for sunlight. Nowhere in the pamphlet did it say that jaundice could lead to brain damage.

The day after Cal returned home, he became lethargic and started to have trouble nursing. Alarmed, Sheridan brought him in to see the pediatrician, but still no test was done and they were sent home yet again. Another day passed and Cal only got worse. Finally, he was admitted to the hospital and started on phototherapy. However, the treatment of Cal's jaundice didn't begin until it was too late. On the sixth day of his life, in his mother's arms, Cal stiffened, arched his neck, and let out a high-pitched cry. Later Sheridan would learn that Cal had displayed all of the classic signs of kernicterus, a condition that occurs when bilirubin gets too high and crosses the blood-brain barrier, leading to severe brain damage. Sheridan literally watched as her baby's brain was being overcome by neurotoxicity that could have been prevented. It was an experience that would haunt her for the rest of her life.

Although rare, kernicterus is devastating. It can result in a range of irreversible neurological damage, and for Cal, the disorder meant that he developed cerebral palsy, hearing loss, crossed eyes, and speech impairment, among other abnormalities, and would need care for the rest of his life. As if that weren't enough to deal with as a new mom, what really ate Sheridan alive was that it hadn't had to be that way. Through a series of tragic lapses in medical care, the urgency of Cal's condition wasn't recognized until the damage was done.

Years later when Sheridan's daughter was born with jaundice, she was quickly tested and successfully treated with phototherapy. When Sheridan saw how easy it could have been to prevent Cal from numerous disabilities, she cried, feeling devastated all over again for her son. But then she got to work. She hit the road to campaign for universal bilirubin screening, something that could be added to the routine newborn care for about a dollar. She spoke at conferences, testified in front of health-care agencies, and formed a nonprofit with other mothers whose kids had kernicterus. To most people who hear Sheridan's story, the answer seems obvious: Do the damn test. Make it mandatory. Of course! But while she was able to make a lot of progress with certain commissions and health-care organizations, she also got a huge amount of pushback from the medical community.

The doctors and heads of committees that make up guidelines for medical screenings were upset that she was trying to change practice based on "emotional stories." Kernicterus was rare enough, they argued, that you shouldn't alarm new parents, who already have so much on their plates. And you shouldn't second-guess doctors. As a patient-safety advocate, Sheridan came up against a number of obstacles that seemed to be more about objections to changing the medical culture than objections to the science. For Sheridan and her son, care reliant on something as subjective as a visual inspection had had disastrous consequences, and she was determined to make sure more kids weren't harmed for lack of a simple screening test. Sheridan's campaign succeeded in putting kernicterus front and center on doctors' radars. She got the Centers for Disease Control to issue an alert to hospitals about the rise in cases, and she convinced the Hospital Corporation of America, a major hospital chain, to require all newborns be tested before discharge. Thanks in part to Sue Sheridan, in 2004 the American Academy of Pediatrics officially recommended that every child have a bilirubin screen in the first twenty-four hours of life.

As someone deeply embedded in medical culture, I know there is often a lot of resistance to changing practice guidelines, and much of that resistance is warranted. That's why before CYW took steps to share our screening tool, we wanted to put some feelers out there and have a good listen. By talking to colleagues, we learned some valuable information about the potential difficulties of implementing a screening protocol for ACEs. There were some thoughtful concerns and questions, but there was also some good old stubborn resistance to implementing another screening protocol.

I get the fact that doctors can't screen for everything. Like most of my colleagues, I find fifteen minutes to be a laughable amount of time to accomplish all of the things a pediatrician has to do in a well-child exam. In that time, we have to look at height and weight, vision and hearing, and growth and development and ask about eating, sleeping, peeing, pooping, screen time, and the dozens of household hazards ranging from peeling lead paint to unsecured firearms. And that's before I even pull out my stethoscope. After about my second patient of the day, I find myself starting every visit with "I'm so sorry you've been waiting."

Our team's response to all of those concerns was to develop a protocol that could be completed in three minutes or less. We recognized that it was important not only to recommend to doctors that they screen, but also to help them understand *why* they should screen, how to screen, and what to do when they detected ACEs. So we decided to pull together a user guide that would go along with the screening tool and answer all those questions.

Around the time that I was seeing Lila, we made our screening protocol available for free online. We knew that getting folks to do things differently was really hard, which was why we set the goal for one thousand downloads over the next three years. To my surprise and our collective delight, over twelve hundred clinics and practitioners in fifteen countries downloaded the tool in just one year, blowing past our goal. When our team reached out to a focus group of doctors who had started screening for ACEs, they all said that they would never go back to not screening. It's like a bell you can't unring.

Building on the positive feedback we got from making our tool available to doctors, we went a step further, creating a network for pediatricians around the country to learn together about how to screen, what to do with a positive screen, and how to advance the care of children with toxic stress *faster*. It's my hope that our National Pediatric Practice Community on ACEs will bring us closer to the day when ACE screening is a universal part of health care. I believe in my core that we *will* get there.

I've seen what a difference early identification and early intervention has made for my patients with ACEs, and like Robert Guthrie and Sue Sheridan, I am on a mission to ensure every kid across the country has the same chance at successful treatment. No matter where the impetus for change comes from — a doctor, a patient, a mother, a tragedy — the important thing is that patients get better care. We have to continue to refine the protocols, catch problems early, and treat our most vulnerable patients with everything we've got.

10

Maximum-Strength Bufferin'

WHEN I WAS A KID, I remember visiting San Francisco from my hometown of Palo Alto, which is only about forty minutes south. We'd come to do all the stuff you do in San Francisco: ride the cable cars, walk across the Golden Gate Bridge, and drive down the crookedest street in the world and up to the tops of the city's famous hills. There is no shortage of posh hilltop neighborhoods in San Francisco, but Pacific Heights is probably the toniest of the bunch.

Sometimes dubbed Specific Whites (for reasons that were self-evident), Pac Heights was totally foreign to the world I grew up in. We knew folks in Palo Alto who were pretty well off, but this was a completely different league. My mom loved driving past the soaring mansions with her car full of children, our faces pressed against the windows. We never dared get out of the vehicle.

I remember that those houses always seemed so quiet. There were no kids playing football in the street the way my brothers and I did on the weekends; no people washing their cars in the driveway. Music wasn't blaring out of windows and there certainly wasn't any FREE furniture on the curb at the end of the month. As a kid, I imagined the people who lived in these fancy houses must be gorgeous, powerful, and totally different than the people I knew.

Fast-forward a couple decades (no need to talk about how many), and I found myself feeling a little like the Fresh Prince of Bel Air. I had married a successful entrepreneur, and for work I was spending an increasing amount of time fundraising for CYW. Both gave me sudden access to the very places and people that had once seemed so mysterious.

Kathleen Kelly Janus was one of those people. I first met Kathleen when she came to visit me at the Bayview clinic in 2012. She had heard good things about the work we were doing in the community. She and her husband, Ted, who had found success managing hedge funds, wanted to learn more. Kathleen had worked for several years at one of the big law firms in San Francisco but devoted so much of her time to pro bono legal work that she ended up leaving to start her own nonprofit. She was a passionate human rights advocate and ultimately ended up teaching law and social entrepreneurship at Stanford. When we first met, I could tell right away that we had the same itch to scratch. Literally. I was thirty-three weeks pregnant, and Kathleen was just a few weeks ahead of me. As we sat across from each other in my crowded office in the Bayview clinic, we both scratched our enormous, itchy bellies.

Over the next several years, as CYW began to take shape, Kathleen and Ted became generous supporters, not just of the work but of me personally. I realized how being in the company of people with big dreams was feeding my resolve to do something similarly big about ACEs. I felt a new type of responsibility to my patients bubble to the surface. I was in the kind of rooms that most of my patients never had the opportunity to set foot in. I knew I had a chance to bring their interests into the rooms with me if I could only figure out a way to get the people with influence to care about them. So when Kathleen told me that she had been going to dinners with other women doing amazing things and invited me to join them, I instantly said yes.

On the night of the dinner, I was running late. My last two patients of the day had needed a little extra time, and as I was circling Kathleen's block looking for parking after making the forty-minute drive across town from Bayview, I was struck by the sense that I might have been in the same city, but it was a totally different world.

Finally, I found a spot that I hoped wasn't blocking Danielle Steel's driveway. Kathleen's house wasn't the biggest house on the block, but it was still pretty impressive. I stepped through the door and into the living room, where everyone was sipping wine or sparkling water and taking in the spectacular view of the bay and Alcatraz that can be seen only from this part of town. Clearly, I was the last to arrive, but no one seemed at all annoyed. Kathleen eventually ushered us into the dining room to take our seats.

Introductions were made all around, and it was immediately apparent that these women were all jaw-droppingly accomplished. One woman was an angel investor and another had worked for the State Department before starting her own international consulting firm with none other than former secretary of state Condoleezza Rice and former secretary of defense Robert Gates. And, this being San Francisco, there were also a couple of successful tech entrepreneurs plus a few women who, like myself, were trying to change the world by starting nonprofits. Prior to the dinner, Kathleen had circulated a feature article on one of the guests, Caroline, that had just run in *Time* magazine. Oh, and there was the minor fact that every single one of them was seriously cover-of-*Vogue* gorgeous. There was just one woman other than me whose hair was not some shade of blond. It was the kind of crowd you could easily love to hate. They just seemed too perfect.

But the minute we started talking it became obvious that I hadn't signed up for dinner with the Stepford wives. These women were trailblazers and they had the battle scars to prove it. We discussed the challenges of running an organization and getting funding; we commiserated about how difficult it could be to "raise the next round" for an idea you believe in deeply. We laughed, we shouted, we talked over one another, we pounded the table. We all shared life hacks, little tips on how to be CEOs and international leaders and high-powered attorneys and still be good mothers and good wives and not lose our freaking minds. At the end of the night, there were hugs and lingering conversations.

Much to my delight, the dinners became a regular thing. The location and the agenda was charged to a different woman each time. When it was my turn to host a few months later, I was excited to put the brain trust on a case.

For the most part, things at work were going well. I had just done a TED Talk that, while terrifying, had been a success. It allowed us at CYW to build the awareness and support we needed to expand our efforts. I had been traveling the country, going everywhere from the Mayo Clinic to Johns Hopkins, talking about toxic stress and the need to screen for ACEs. While the message was clearly hitting home, I continued to see one particularly vexing issue: media reports invariably presented toxic stress as if it happened only in poor neighborhoods. I had set up a Google alert for *toxic stress,* and the title of every article I received contained some version of "the toxic stress of poverty." It was driving me crazy. While I knew all too well that poor communities experienced higher doses of adversity and had fewer resources to deal with it, I was worried that the issue was being framed as a "poor-people problem" or a "black-and-brown problem." I repeatedly called out Dr. Felitti's demographic: 70 percent college-educated and 70 percent Caucasian. But that was not the thing people took away.

On the night of the dinner at our house, my husband, Arno, helped me cook a gorgeous meal. In this case, "helped" means that I chopped things precisely the way he told me to while he put it all together into something resembling a *Bon Appétit* magazine cover. While he whipped and whisked, I told him my plan. I was going to share my frustration with the group and see if they had any ideas.

There was a cold tomato and cucumber soup, a perfectly roasted chicken, and a late-summer salad. As the Pinot Noir began to flow, I laid out my problem to the group. I told them that we were getting traction with spreading the message, but it seemed as if the world was missing the point that toxic stress is about basic human biology and that adversity happens everywhere, among all races and geographic areas. I shared my fear that if this became a poor-people-of-color issue, we would miss an opportunity to help all children. I asked how they thought I could get my colleagues to understand that it was important to screen *everyone* for ACEs, not just people in low-income or vulnerable communities.

For a handful of seconds, everyone was silent. But before I could worry that they had no idea what I was talking about or, worse, that something had gone terribly wrong with the soup, they all started talking at once. I had put the question out to a bunch of women rock stars, but they answered as mothers, wives, and daughters.

Kara, the angel investor, jumped in. "I think the problem is that it's just so behind the scenes in other communities. I mean, my dad was an alcoholic, and it was really brutal. But he could hold down a job, so no one knew."

Heads nodded.

As the conversation popcorned around the table, fully half of the ten women shared their own histories of some significant ACEs. Much of it was very similar to what I heard from my patients in Bayview — parental mental illness or addiction, sexual assault, physical or emotional abuse, domestic violence — but what struck me was how hidden it had been. Looking at these women, at what they had accomplished, at the lives that they had built for themselves, no one would have guessed that half of them had experienced some pretty major adversity as kids.

Ultimately, Kara piped back up. "I guess the big question is, what can you do if you know you have an ACE score? I mean, is knowing really going to make a difference?"

I was about to launch into my standard response, but before I could, I heard Caroline let out a sigh and put down her spoon. Even more than her Scandinavian-supermodel looks, Caroline's bearing is what I find remarkable about her. She may be the most analytical person I have ever met. Her brain is like a computer. No matter what the question is, when Caroline responds, you get the sense that she has calculated all of the options and is responding with the solution that has at least a 99.4 percent likelihood of success. Yet suddenly, something in her face — in her very demeanor — changed. Everyone looked her way.

"Oh, you guys," she said, shaking her head, "it makes all the difference in the world."

While the salad was served, Caroline told us her story.

She had met her husband while she was in graduate school at Stanford. Both artsy and math-y in equal measure, she had degrees in art and computer science and was totally fascinated by the symbiosis of man and machine. Naturally, she fit right in with the scores of people whose life work in the 1990s was finding patterns in the huge data sets being generated by this new thing called the Internet. It seemed obvious to Caroline that a visual tool was needed, so she spearheaded the development of software that helped researchers visualize information in a way that allowed them to more easily compare trends in data.

The software was a huge success and launched Caroline's career. So she dropped out of Stanford and started a company to develop and license the software. It was through that work that she met a man named Nick, who was tall, handsome, and super-intense.

Caroline was drawn to Nick's passion for politics and science and loved how he could wax philosophical for hours about what he saw as the inevitable future in which artificial intelligence saves the world. Things moved quickly, and within a few months they were living together. Soon they were married, and for the most part it was wonderful, but after a few years Caroline felt that something was off. Something didn't feel quite right, but she couldn't put her finger on it.

So when she found out she was pregnant, that first moment of realization wasn't what Caroline had always imagined it would be. She didn't squeal and rush to tell Nick. In fact, she thought about not telling him at all — she even considered leaving before her pregnancy began to show, breaking it off with Nick and moving out. This urge felt like both a betrayal and somehow the right thing to do. Still in her twenties, Caroline had started a company and was on the way to some serious success. This was where her life was, and besides, she did love Nick. When things were great, they were *really* great. It just didn't feel like things had been that great lately.

When Caroline told Nick about the baby, he was sweet and excited. During her pregnancy he would rub her belly as they lay in bed and say, "Just imagine, a little boy for me to build robots with." He helped her lever her enormous belly out of chairs and brought her water to make sure she stayed hydrated.

But after Karl was born, things changed. Nick quickly became frustrated that Caroline was giving all her energy and attention to the baby. As most moms know, a newborn is a giant well of need at the center of your world. Everything else comes second, third, or not at all. Caroline understood this change of regime had to be hard for her husband. He'd been dethroned pretty rapidly. Gone were the days of her tousling his hair as she walked by the couch on her way to whip up some dinner. Instead, she felt frazzled and overwhelmed, and more often than she cared to admit, it just seemed like he was getting in her way, making her mom-job harder than it needed to be. Soon, the smallest things became epic arguments.

Nick's drinking escalated dramatically after the baby was born. He had always been a partier, but after Karl arrived, things went off a cliff. Soon he was having issues at work and was fired from a string of jobs. As the months went by, it seemed to Caroline that she spent more time figuring out how to avoid fights with Nick than enjoying his company. Everything set him off. He refused to help take care of Karl, so when she went back to work, they got a full-time nanny. He resented when Caroline talked on the phone with her dad, and when she finally found a moment to get out of the house and have lunch with a girlfriend, Nick was furious.

Despite how angry he constantly was at her, Nick didn't want Caroline to be very far from him. At first he just moped when she spent solo time with friends and family, but it wasn't long before he was giving her ultimatums ("It's them or me!"). Eventually, Caroline decided it was easier to avoid the drama altogether and just stay home and watch TV with Nick. He seemed at least a little happier that way. She found herself making excuses to her girlfriends for why she couldn't go out.

One evening when Karl was about six months old, Caroline and Nick were in the kitchen making dinner when something happened that set him off. Like a discarded cigarette igniting a wildfire, it was a small transgression with big consequences. Years later she wouldn't even remember what it was, but she would never forget the sound of Nick screaming at the top of his lungs, slamming cabinet doors. Caroline shrank into silence. She knew better than to try to argue, and for about thirty seconds after he stopped yelling, the entire kitchen was quiet. Then from the breakfast nook, Karl let out a wail. When Caroline looked at her son, his face was beet red and he was letting out the shrieking, gasping cries that rips at any mother's heart. Still frozen, Caroline thought to herself that she had never heard that particular cry from him before. Just then, the nanny swooped in, scooped Karl up, and took him into the other room.

Caroline wondered how the hell she had gotten to this place. On the surface, things appeared to be going well. Her company had been acquired and she had joined the leadership team of one of the biggest firms in Silicon Valley. But at home, things were awful. The sound of the garage door opening, announcing Nick's arrival, would set her heart pounding, and when she heard his keys jangling at the front door, she would brace herself for what might be coming next. She was a smart woman. After all, she was managing hundreds of engineers and computer scientists every day. She knew there had to be a way for her to manage this situation. She just hadn't figured it out yet.

In the rare moments of connection and tenderness with Nick, she would gently ask him why they fought so much. "This can't be normal, can it?" He had one of two reactions whenever she brought up the possibility that something was wrong. When a bad mood was barely below the surface, he would go off on a diatribe about how all her friends were against him. He'd say they were just jealous because he and Caroline loved each other so much while their own marriages were boring and passionless. If he was feeling playful, he would tease her about being a "typical woman." He'd compliment her, saying that she was too smart to get caught up in some romantic-comedy-induced delusion of the perfect relationship. He'd call her babe and say this was just the way love worked in the real world; you laughed *and* you screamed sometimes. Either way, you knew the other person loved you, so you gutted it out.

Shortly after Karl turned three, the family moved from the center of town to a new home, a big house that was as secluded as it was beautiful. The live-in nanny who had been taking care of Karl since he was born was unable to come with them. Up until that point, Karl had been a confident, happy kid. He would run up to strangers on the street and exuberantly shout, *"Hi, I'm Karl!"* After the move, Caroline noticed that Karl became withdrawn and shy. Soon they were getting calls from his nursery school. His teachers complained that he had started hitting other kids in his class. By his fourth birthday, the school had had enough. They insisted that Caroline and Nick take Karl to be evaluated for ADHD.

Caroline was worried. In addition to his short fuse at school, she noticed that Karl had become quick to cry and tantrum at home. More concerning, he was suddenly getting sick all the time. He had always been a healthy kid (she had breastfed forever), but lately he constantly had a cold or a tummy ache or a headache. She wondered if the new house was too damp.

They were referred by their pediatrician to a top clinic for ADHD assessment where Karl was seen by a seasoned clinician. He evaluated Karl and his parents together, then spent some time with Karl alone. While the four-year-old played tentatively with one of the medical assistants nearby, the doctor told Caroline and Nick what he had observed.

"Look, this is going to be hard to hear, but your child is lacking the protections of childhood," he said.

"What does that mean?" Caroline asked.

"He's being exposed to psychological trauma. He needs a more peaceful, less stressful environment. We believe that is what's contributing to his ADHD."

For Caroline, the part of the conversation that would haunt her later was also the part Nick was unable to accept. *Exposed to psychological trauma.* That's what the doctor had said, but Nick ignored everything but the term *ADHD,* and while he was great about making sure Karl got his Ritalin, he told her the rest of what the doctor said was bullshit.

While some of Karl's teachers were happy that his behavior was more manageable, Caroline was disturbed that her son seemed "totally zombified." Gone was her spirited, willful child and instead here was a glassy-eyed kid who couldn't eat because the medicine gave him stomach problems. They tried a couple of different medications, finally ending up on Adderall, but Karl still hated the way it made him feel. In school, he was calmer, but Caroline worried that he wasn't learning.

When Caroline started having what she thought were panic attacks in the middle of the night, she began to wonder if maybe *she* was the problem. Maybe this sleepless-heart-pounding thing wasn't about Nick or their relationship; maybe it was just her. Was she working too much? Did she have some kind of condition? She didn't know, but knew she needed to fix it, so she went to therapy to try and figure it out. The doctor she saw prescribed exercise and time to herself. *Yeah, right.* She laughed. By this time, she was running one company and consulting for another. But the doctor was serious, telling her to book "Caroline-time" into her schedule just like she would a marketing meeting. He said that she'd be accountable to him for that time — he would check in with her about whether or not she kept the meeting with herself. For a while, she tried, dutifully putting it on her calendar, but it didn't work. She'd hijack the time, use it to finish that one project that simply couldn't wait. This went on for months before her boss finally intervened.

"Why don't you use my personal trainer?" he suggested. "I insist."

When she saw her boss's face, it dawned on her that perhaps she hadn't been hiding her personal stress as well as she had thought. Caroline knew enough to accept his offer.

With her boss's support, Caroline found that it was easier than she had anticipated to squeeze some yoga into her schedule between meetings. Somewhere between tree pose and downward-facing dog she began to feel her stress slowly lifting in waves. For a while, she woke up less and less in the middle of the night. But it wasn't long before her me-time became an issue for Nick, sparking an epic fight about her selfishness. It didn't matter to him that she was busting her butt as the sole breadwinner for the family; he thought she should spend less time working and more time with Karl and him and that she should definitely, *definitely,* not be taking time away from the family just so she could try to *look good.* He began publicly posting his opinions about her online.

Caroline felt like a fly stuck in amber. Nothing she said or did could get Nick to change his behavior. She knew that his rage was terribly harmful for Karl, but she told herself that, after all, Nick had never hit Karl, or her, for that matter. She resolved to make sure that Karl would never be alone in Nick's care. Divorce meant shared custody, and she was panicked at the thought of not being there when Karl spent time with his father. What if he got drunk and drove with Karl? What if he flew off the handle and screamed at him? As miserable as she felt, it wasn't about her. Caroline needed to be there for her son, so she would stick it out, come hell or high water. So nothing changed. And maybe it wouldn't have if it hadn't been for the unimaginable courage of her seven-year-old son.

One day, during a typical blowout, instead of retreating to his bedroom like he usually did when his parents fought, Karl stood in the doorway and watched as his dad berated his mom. When it was over and his father had left, Karl went to his mother and took her face in his hands.

"Mom," he said, looking her straight in the eye, "we have to leave."

. . .

Two years later, Caroline sat in a darkened room with six other women watching a video. They were all strangers to her, other moms who had also filed restraining orders, other women who she imagined were just as surprised as she was to see themselves reflected in a low-budget, court-mandated video. But the video wasn't about the women; it was about their children. A couple argues in a bedroom upstairs while a little girl stares blankly at a TV. A little boy is nonresponsive when a teacher asks him questions at school. Another boy lashes out at his sister, hitting her just like he's seen his dad hit his mom. Caroline remembered thinking as she watched that the video's message was what you would expect — witnessing physical abuse is obviously bad for kids, everybody knew that. But what made her sit on the edge of her seat and her hands go numb was what the video had to say about verbal and emotional abuse.

It was just as bad for kids, and in some ways, worse.

The video showed kids with symptoms just like Karl's. But it was when they showed a baby starting to cry as his parents argued that Caroline remembered Karl wailing in his high chair.

She began to weep.

. . .

Years later at my dining-room table, Caroline's tears were gone, but her astonishment wasn't.

"Fifteen years, I lived like that," she said, shaking her head, "and I thought it was normal. I blamed myself. I thought something was wrong with me all those years. I wish someone had shown me that video when I was in high school."

As Caroline wrapped up her story, faces around the table revealed a mix of empathy, solidarity, and utter disbelief. Despite the fact that many of the women at dinner that night had known Caroline for years, none of them had heard that story before.

She told us that it wasn't until her lawyer actually said it that she ever contemplated that what had happened could be considered emotional abuse. The yelling, the intimidation, and the controlling behavior — all of a sudden, she could see it for what it was.

"How is Karl doing now?" Kathleen asked.

"So much better," replied Caroline.

She told us that not long after they moved out, she began to see a change. Karl wasn't so quick to get upset, and he just seemed calmer in general. She took him back to the psychologist, and she and Karl were now in therapy, both together and separately. But ironically, the things that seemed to make the biggest difference for Karl were the changes she made for *herself*. Caroline created more time for her son and herself. She rediscovered her love of drawing and painting and ballet. She found herself able to slow down and open up. She described feeling calmer and gentler. Karl completely fed off his mother's energy. Together, they took up rock climbing and began doing yoga poses in the living room of their new apartment. Eventually, they decided he should stop the medication for ADHD.

Initially, when Karl went off the medication, some of his problem behaviors returned. He was very reactive and would get upset quickly. Caroline spent time helping his teachers understand how to work with him. They made sure that he was actually writing things down, that he switched his attention from one task to the next intentionally and then came back to the original point. For years, he had missed out on learning those basic skills because he was so subdued. From then on, when the challenging behaviors came back, Caroline, his teachers, and his therapist were able to address them in partnership successfully.

"Honestly, it sounds like Karl was experiencing toxic stress," I said. "It makes perfect sense that he did so much better because exactly what you did *is* the treatment for toxic stress. Number one, reduce the dose of adversity; number two, strengthen the ability of the caregiver to be a healthy buffer. Your getting healthy was actually an incredibly important part of the equation. It's like when a flight attendant tells you to put your own oxygen mask on before putting it on your child. That's no joke. Your stress response was dysregulated, which made it impossible to help him regulate his. That's the mechanism that is so critical to understand. You getting out and taking care of yourself wasn't selfish — it was the exact right thing to do for Karl."

Caroline nodded. "I've noticed that the more I do for myself, the better he copes with things."

"It's crazy how resilient kids can be when they have a strong buffer," I said.

"It's true. Now when he has supervised visits with his dad, he'll come back and something will have happened to bother him, right? He might have a shorter fuse for a couple of days, but after a few more days of just doing our usual thing, he's back on track. I just wish I had known that earlier," Caroline said, shaking her head. "I would have gotten out a lot sooner."

"I see it in my patients every day, and, girl, it is *rough*. I'm so sorry you had to go through that," I said. "Situations like yours are exactly why we need to do screening for *everyone*. Because most pediatricians, if they saw you, Ms. Gorgeous-*Time*-Magazine, roll into their exam room, they wouldn't ask about potential adversity at home. They might be afraid of offending you or assume that because you're so well put together, nothing like that could possibly be happening in your home. But if screening is part of a protocol that they do as a matter of course, they'll be able to identify what's going on."

Janet, a dynamo who runs a successful online retailing business, chimed in from the other end of the table, "Okay, so can we get real here for a minute? It's obvious why screening every child is a must, but what do you do if you're an adult and you had ACEs as a kid? Is there treatment for that? Honestly, I am thinking about my husband, Josh, right now."

"Absolutely," I said. "It's never too late to start rewiring your stress response."

"The impact of interventions for toxic stress may not be quite as dramatic in adults as it is in our kids, but it still can make a big difference. This might sound simple, but I cannot overstate this: *The sin-*

gle most important thing is recognizing what the problem is in the first place."

I shared with them my observation that many people with overactive stress responses don't know what's happening in their bodies, so they spend all this time chasing down the symptoms instead of getting to the source of the problem. Once folks understand what's going on, they've taken the first step toward healing. I went on to explain that for toxic stress, the six things that I recommend for my patients — sleep, exercise, nutrition, mindfulness, mental health, and healthy relationships — were just as important for adults. Checking in on how you are doing in those six areas and talking to your doctor is a good place to start. If necessary, you can request a referral to a sleep specialist, a nutritionist, or a mental-health provider.

The other important piece I mentioned was that adults with high ACEs were at increased risk of health problems, which was why it was important for them to ask their doctors if they had heard of the ACE Study. A doctor can help you understand how your ACE score and your family history affect your risk for certain illnesses, and then the two of you can work together to create a plan for prevention and early detection. The great news is that there is now a field, called integrative medicine, that is dedicated to looking at the whole person and using the latest science to improve health and well-being. The cool thing about integrative medicine is that it's interdisciplinary, just like our team at CYW.

There are lots of different ways to combat toxic stress. If you hate yoga and rock climbing, you might be into running or swimming. That's fine; as long as you're doing some kind of regular exercise for about an hour a day, that's what matters. Likewise, there are lots of types of mental-health interventions that work, but the most important thing is to make sure that they are trauma-focused. Ideally, you want to maximize all six of those things, especially for adults, because our brains aren't as plastic as they were when we were kids. But the general idea is, the more of the six things you do, the more you'll reduce stress hormones, reduce inflammation, enhance neuroplasticity, and delay cellular aging.

"Of course, it's also a good idea to cut out the stuff that accelerates inflammation and cellular aging, like cigarettes, and to minimize neu-rotoxins like alcohol," I said, tapping on my wineglass.

"All the fun stuff, is what Josh would say," said Janet, smiling.

"Well, when you tell him that if he cuts down on the beer, he can ramp up the intimacy, he might not mind as much," I said.

"Does that fall under the exercise category?" asked Janet.

I laughed.

"There's that, but it has more to do with the healthy-relationships piece of things. Sometimes I think folks out there are waiting for a fancy pill to show up and they're missing the point that we, as humans, have a profound power to heal ourselves and one another. Look, the research defines *toxic stress* in children as long-term changes to brains and bodies in the absence of a buffering caregiver. So think about the flip side of that for us adults. We can damage each other's health by re-peatedly activating the stress response, but we also have the power to heal ourselves and others biologically. Let me give you an example — any of y'all get the drug Pitocin when you were having your babies?"

Heads nod.

"Well, that same drug, oxytocin, is actually naturally produced by our bodies. It's released in huge amounts during childbirth and it not only helps the uterus contract to push out the baby, it's also this in-credibly powerful bonding hormone, so that when your baby comes out, you've never seen anything so beautiful in your entire life and you would take a bullet for this little cutie. And oxytocin isn't released only during childbirth; it's also released during sex and with hugs and snug-gles and healthy relationships. And it buffers the stress response by ac-tually inhibiting the HPA axis — the brain and body's stress-response circuitry. Plus, it has been shown to have antidepressant effects. We lit-erally have the capacity to change our own and one another's biology. We don't need to wait for a pill. I honestly believe that, right now, we have some very powerful tools to interrupt the intergenerational cycle of ACEs."

"Do you think your ex-husband had ACEs, Caroline?" asked Kath-leen.

"Absolutely."

She went on to tell us that Nick grew up in a well-to-do suburb in

Connecticut. His father was a doctor and his mother was a respected engineer. But Nick's household was less like the Huxtables' he grew up watching on TV and more like a scene from Whitney Houston and Bobby Brown's short-lived reality show. Nick's dad had a pretty significant problem with cocaine and marijuana. Then Nick's parents divorced when he was ten and he endured a series of stepmoms, each with a progressively escalating coke habit. For the most part, Nick's dad was able to fly under the radar, functioning as a physician for years without any major incidents. Home, however, was something else entirely. Nick's dad and his various stepmoms would get into heated, drug-fueled altercations. Nick always used the same word when he described his dad's house — *crazy*.

"Oh my goodness. That's so sad," I said. "The thing about it that breaks my heart is that we know that most ACEs are handed down from generation to generation. If Nick had figured out that what he experienced was ACEs and that he probably had a dysregulated stress response that he needed to deal with, can you imagine how differently things could have turned out for you and for Karl?"

"It is ridiculous that *everyone* does not know this. How do we get people to pay attention to this as something that affects someone they love whether they know it or not?" asked Janet.

"That's what *you're* supposed to tell *me!*"

"Well, for starters I think Caroline should call up *Time* magazine and tell them they've got their next cover story," said Kathleen.

After that, everyone started talking at once. The conversation leaped from what everyone's version of "normal" was in her own childhood to ideas about how to change the status quo by improving awareness and education around ACEs. The night was a total success, but not necessarily because I had gotten some practical "get the word out" strategies (though I definitely did). The evening showed me the power of the ACEs framework to open a dialogue about topics that feel largely taboo in our society. I knew statistically that it was likely that I was surrounded by people with ACEs, but I had never had such an open conversation about ACEs outside of the Bayview clinic until that night.

I've often remarked, only half joking, that the biggest difference between Bayview and Pacific Heights is that in Bayview, people actually know who the molesting uncle is. And it's not because the 94115 zip code has a magical force field that excludes anyone who might somehow harm a child or who is experiencing substance dependence or mental illness. These things just aren't talked about.

When I later asked Caroline why she thought there was so much secrecy in upper-income circles, she responded that she believed it was because the risk to reputations was so high.

"We are expected to be perfect. We are supposed to have it all together. The hiding is pervasive because exposure can cost people their careers. By the mere fact that we are hiding it, we are perpetuating it."

. . .

After that dinner, it became clear to me that these hidden ACEs were hindering not only the people experiencing them but also the movement that CYW was trying to catalyze by perpetuating the myth that adversity was a problem for only certain communities. Caroline's bravery in sharing her story moved me to my core. ACEs and toxic stress thrive on secrecy and shame, both at the individual level and at the societal level. We can't treat what we refuse to see. By screening for ACEs, doctors are acknowledging that they exist. By being open about ACEs with friends and family, people are normalizing adversity as a part of the human story and toxic stress as a part of our biology that we can do something about.

Toxic stress is a result of a disruption to the stress response. This is a fundamental biological mechanism, not a money problem or a neighborhood problem or a character problem. That means we can look at one another differently. We can see one another as humans with different experiences that have triggered *the same physiological response.* We can leave the blame and shame out of it and just tackle the problem the same way we would treat any other health condition. We can see this problem for what it really is, a public-health crisis that is as indiscriminate as influenza or Zika.

I closed the door after my last guest left and sat down at the table where we had all been gathered moments before. I realized that something important had just taken place. After years as an unwitting detective in Bayview and Pac Heights and a bunch of places in between, I had finally figured out what I needed to do to create a sea change in the fight against ACEs and toxic stress. I had inspected all the wells in all the towns and discovered that not only were they deeper than I had ever imagined but, more important, *they were all connected.*

IV

Revolution

11

The Rising Tide

THE DINNER WITH CAROLINE seemed to jump-start a surreally good streak for my shout-it-from-the-rooftops campaign regarding ACE impacts and treatments. The American Academy of Pediatrics invited me to keynote their first-ever national conference on toxic stress, and I was even invited to the White House to give a briefing for leaders of eight White House agencies. It was a serious pinch-me-I-can't-believe-this-is-happening moment.

And I wasn't the only one talking about ACEs. More and more I was hearing leading voices calling out the need to identify and address the impacts of toxic stress. When I had the opportunity to visit the National Institutes of Health, Dr. Alan Guttmacher, head of the National Institute of Child Health and Human Development, mentioned that he had actually seen my TED Talk and shared with me his belief that "the developmental origins of disease are the future of medicine." This resulted in the rarest of responses from me—utter speechlessness. How ACEs affect biology was suddenly a topic for discussion, even in circles where those conversations hadn't previously been happening.

So when I began my talk on the need for ACE screening at a New York City conference in the summer of 2016, I was certain the cross-sector group of scientists, activists, educators, and policy wonks would be the perfect partners for thought-storming ways to make universal ACE screening a reality for all children. The one difficulty was that since I'd recently given birth to my youngest son, my body had become the milky equivalent of Old Faithful. There had been a whole day's worth of talks after mine that I couldn't tear myself away from, so by the time the moderator kicked off the closing discussions, I was a hurting mama. I had to sprint to the lactation room to pump.

Nearly an hour passed before I finally returned with seven ounces of liquid gold for baby Gray (or Grayboo, as I had taken to calling him as soon as we met). I'd been hoping to catch at least some of the wrap-up Q&A, but the woman before me in the lactation room had taken her sweet time. As I scooted into the back of the conference room, squeezing between chairs and whispering *excuse me*s, I picked up on a weird vibe. The air was thick with the feeling you get when something has gone sideways — and I had a sinking feeling it might have something to do with me. I'd come in at the tail end of someone's comment and registered only the tone, which was decidedly tense. After that, the conference organizer got up, thanked everyone, and closed out the day.

What the heck did I miss?

I packed up my things and was making my way to the wine-and-cheese portion of the agenda when I was stopped by Jeannette Pai-Espinosa. Though petite in stature, Jeannette has a big presence. Growing up in Kansas City as the daughter of South Korean immigrants, she has the confidence of someone who has survived her fair share of storms and as a result knows how to navigate the world better than most. She had walked up to me with an expression on her face that said, *Don't worry, girlfriend, I've got your back.* Although we had never met, I knew Jeannette by reputation. She was the president of the National Crittenton Foundation, an organization that works in thirty-one states and the District of Columbia to support the self-empowerment of young women and girls. The National Crittenton Foundation had landed on my radar because they had adopted a mandate to address root causes of poor outcomes for girls, and in doing so, put ACEs at the core of their work. I had heard that the foundation's ACE-informed approach to breaking intergenerational cycles of poverty, poor outcomes, and violence was yielding powerful results. Jeannette was a fellow foot soldier who witnessed, day in and day out, the true impact of childhood adversity.

Jeannette skipped the handshake and wrapped me in a hug.

"Well, *that* was interesting!" she said as she stepped back.

"I just got back from pumping — what the heck happened?" I asked.

"People are upset! There was a big conversation about why it's dangerous to screen for ACEs because it'll be used to label low-income children of color as 'brain-damaged,'" Jeannette replied, shaking her head. "Which is nuts, because none of the folks who were raising these concerns are actually screening for ACEs."

"What the heck?" I was crestfallen. "Did they not hear me say that this happens in every community? This is about basic biology."

"There's a lot of misunderstanding" came a voice behind us. I turned around and recognized Nancy Mannix, the chair of a foundation taking on ACEs in Alberta, Canada. Nancy had every bit the look of a foundation patron, wearing a gorgeously tailored cream-colored suit and rocking a dark brown bob that reminded me of Jackie O. Earlier in the day I had heard Nancy stand up and share her experience with bringing the brain science and ACE screening to decision-makers and practitioners across the province. Listening to Nancy, I had been seriously impressed by her insights on the ground game. It was clear that she wasn't afraid to roll up her sleeves and get her hands dirty. I had made a mental note to connect with her, so I was thrilled when she approached Jeannette and me. "We saw the same thing when we were bringing ACE screening to Alberta. The greatest pushback comes from the folks who don't know the science and aren't doing it. I've never heard someone say, 'I tried screening, but it didn't work' or 'We had to stop.'"

It took only a few minutes for Nancy and Jeannette to fill me in. It turns out that as part of the summary of the day's talks, my call for universal ACE screening came up and was met with some pretty fierce criticism once the floor was opened up for comment. The most passionate resistance came from a few people who felt I was "medicalizing" adversity when they, as community activists, had spent a long time trying to solve the inequities giving rise to it. The loaded term *biological determinism* was even thrown out there.

These criticisms stung for a couple of reasons but primarily due to the fact that I had spent my entire career working shoulder to shoulder with community partners to improve the health of vulnerable children. That's what had driven me to understand ACEs and toxic stress in the first place. Somehow, all of that was missed and I was being painted as "that doctor from San Francisco telling us that our kids are brain-damaged." I felt as confused and disoriented as I had when I first heard Sister J warn about the "toxic dust" at our site.

"I get the concern about labeling, trust me, but it's just not the reality," Jeannette said.

She had firsthand experience of what could happen when ACE screening was deployed on a large scale. Across the diverse agencies supported by Crittenton, whether it was a child welfare agency or a juvenile justice organization or a group serving young moms or sex-trafficking survivors, Jeannette had seen that the information about ACEs empowered and truly transformed young women; it didn't label them.

She told us a story of a recent trip when she was accompanying eighteen women and girls from various Crittenton programs located in eighteen different states to Washington, DC, to educate policymakers about ACE screening. As she was presenting the data, Jeannette said, there was a woman sitting right in front of her who lowered her head and began sobbing. Jeannette remembered thinking to herself, *This is the one time someone is actually being triggered by this.* She had never, ever seen that. She stopped the meeting, told everybody to take a break, walked over to the young woman, and sat down with her.

"Are you okay?" she asked gently.

The woman shook her head. "Oh, no. These are not . . . I'm not upset. You didn't upset me."

Jeannette leaned in, confused.

The young woman continued, "These are tears of pure, unadulterated joy."

"Why joy?" Jeannette asked.

"Because I understand now why I am this way. I understand why my siblings are this way. I understand why my mother raised us the way she did. I understand that I can break this cycle for my children and I understand that I'm not a victim, I'm a survivor."

Since that day, Jeannette told us, this young woman had begun reading everything she could on ACEs and toxic stress. And though she knew it would be a long struggle, she said, "I understand that I got here, that my family got here, over generations. And it will take me a while to fully process all that. But I know that I can make better choices. And not just for me. I can stop my children from having a score of eight, nine, or ten." The young woman had scored a ten out of ten on the ACE screen.

Over the years, the National Crittenton Foundation has found the ACE scores to be one of their biggest tools for self-empowerment and advocacy. Once the women they support have the information, they are able to look at the context of their lives differently. Then they no longer feel they are to blame or that they're stupid or that there's something wrong with them. Once they understand how what happened in the *past* can affect how they feel in the *present*, how they see themselves and their healing process changes. They understand that their bodies have experienced a normal reaction to abnormal circumstances across the span of their lives. Many times, they'll call their siblings and say, "This is it, this is what's been going on with us!" The older girls in the Crittenton programs began talking to the younger girls about ACEs and how they were affected simply because they wished someone had told them.

. . .

As we got deeper into our conversation, Nancy Mannix shared more about her experience in Canada. She had spoken to the criticism concerning overmedicalization. A few people were resistant to the idea that toxic stress was a physiological problem in the first place, suggesting that ACEs and their impacts were simply normal human or cultural problems that had no business being met with a medical diagnosis, so why not leave the learning problems to the teachers and the behavior problems to the therapists? The expressed concern was "overreliance on neuroscience."

Nancy's experience in Alberta made her an ardent believer in the science of toxic stress and in routine ACE screening as a critical part of regular medical care. In 2005, she had stumbled on Felitti and Anda's research while trying to understand the role of childhood trauma in addiction treatment. Around the same time, Nancy also discovered the work of the Harvard Center on the Developing Child, which clarified for her the scientific basis for using ACEs to assess for toxic stress. At the time, her job was to identify individuals and organizations doing important work in the fields of child development, mental health, and addiction. When she first read the ACE Study, she experienced a jolt of understanding of the deep connections between each of the fields she was passionate about.

At the time, Mannix and her team observed that most addiction treatment was grounded in the belief that clinical work should focus on the patient's future, which meant clinicians didn't want to spend too much time on their patients' past. Interventions were disparate and based on individual diagnoses. The systems that were supposed to help patients heal were fragmented. Mannix recalled the case of a seventeen-year-old girl with an eating disorder and a cocaine addiction who was sexually acting out. The fact that these behaviors might all be the symptoms of a single underlying root cause wasn't on anyone's radar screen. So she was sent to rehab for the drug problem, sent to a separate clinic for the eating disorder, and "counseled" about the dangers of risky sex. No one realized that the severe adversity the young woman had experienced as a child might be driving her symptoms, and none of the interventions were particularly effective. Mannix and her team set out to change all that.

They began by bringing a group of addiction-treatment providers together with patients to talk about how the system could better serve clients. Some providers were receptive, but others pushed back defensively, insisting they were the experts and that they provided excellent care—these patients were simply failing treatment.

So, Mannix made it her mission to bring the science of ACEs to Alberta. She held what she called an initial "catalytic convening" in the town of Red Deer, inviting clinicians, academics, policymakers, and education experts. She recruited the leading experts in the field of toxic stress to lay out the latest science and create a straightforward and understandable story to explain the impact of early adversity on brain development. This convening launched a multiyear strategy to bring decision-makers and practitioners together with scientists to understand ACEs and the emerging science.

As part of this process, researchers at the University of Calgary launched a study, recruiting over four thousand patients from primary-care clinics and asking about ACEs as well as health status and mental-health measures. Much like the original ACE Study, the population was 83 percent Caucasian and 82 percent college-educated. What the researchers found was that the numbers fell within a few percentage points of Felitti and Anda's results — demonstrating that Alberta was as affected by ACEs as anywhere else. People with high ACEs were (again) shown to be at much higher risk for depression and anxiety and also to have a greater risk of asthma, autoimmune disease, food allergies, cardiac disease, chronic obstructive pulmonary disease (COPD), migraines, fibromyalgia, reflux disease, chronic bronchitis, stomach ulcers — and the list goes on.

People were astounded to see the profound effects of ACEs that had previously been unrecognized in their communities. After they got over their shock, they came together to find solutions. Doctors and health programs began regularly screening for ACEs in both outpatient clinics and inpatients, and policymakers put forth contract requirements with agencies receiving government funding to be competent in the brain science. The Alberta Family Wellness Initiative, as it would come to be called, made its mark in Canada by turning "what we know" about early adversity and health into "what we do" in practice and service delivery. So on this day, Nancy Mannix was eager to rebut the "overreliance-on-neuroscience" bias, eager to proselytize for competency in the science and routine ACE screening, and eager to insist on mobilizing the powers that be in support of better systems to create better care.

Jeannette and Nancy and I had all come via different paths, but we had arrived in the same place and were focusing on the same source of the problem. Standing with them, I could feel the beginnings of a true public-health response coming together.

. . .

But the day's contentious conversation had illuminated yet another point of resistance. While I had expressed my opinion that primary-care clinics were the ideal place for ACE screening, I had also said that enough kids had been sent to my clinic by teachers requesting a diagnosis of ADHD and medications that I knew that the doctor's office wasn't the only place that needed fundamental understanding of toxic stress. This statement opened a hornet's nest: one woman in particular wondered, as I heard later, whether ACE screening in schools could be used to label low-income kids and stigmatize them even further.

Whenever I had a question concerning ACEs and education, I knew whom to turn to — fellow physician and ACE trailblazer Dr. Pamela Cantor. Her organization Turnaround for Children was leading the charge to bring the science of ACEs and toxic stress into schools.

Turnaround has been at it for over a decade, but Dr. Cantor herself has been working with kids affected by ACEs for a lot longer than that. A psychiatrist by training, she specialized in child mental health and gravitated toward treating kids exposed to trauma. She had developed what she called a Robin Hood practice — as a member of the faculty at Cornell Medical School, she practiced on the Upper East Side of Manhattan and in the South Bronx. Working at one clinic paid the bills so she could work at the other. Unsurprisingly (to me, anyway), the common thread she saw between her work in both communities was exposure to ACEs. Over the years she became more and more involved in research and advocacy that focused on the developmental disruptions caused by trauma. Which is why on September 11, 2001, when the most acute trauma the United States had experienced since Pearl Harbor hit, New York City came calling.

Dr. Cantor was asked to co-chair a partnership commissioned by the city's department of education and help launch a study to investigate the traumatic effects of 9/11 on New York City's public-school children. The partnership worked with researchers from the Columbia University Mailman School of Public Health, and together they undertook what was at the time the largest epidemiological study of an urban public-education system from a mental-health perspective. The commonsense hypothesis going into the study was that the kids in schools closest to Ground Zero would be the most affected and would naturally need the most help.

The data came in the form of maps on huge sheets of tracing paper that the research team could overlay to see the alignment between trauma symptoms and the various neighborhoods relative to Ground Zero. As they lay sheet after sheet over each other, the team found that the data showed a totally different picture than any of them had expected. The distribution of trauma symptoms was not clustered around Ground Zero, which were largely middle-class neighborhoods. Instead, the greatest groupings of trauma symptoms corresponded strikingly with the communities of deepest poverty. The next page of the map revealed that the areas that were most affected were also the communities that had the fewest resources.

Dr. Cantor's response to the data was to get out into the schools and meet the actual children represented by these dots on a map. The first place she visited was an elementary school in Washington Heights, a neighborhood on the border of Harlem.

As she entered the school, Dr. Cantor noted that the hallway to the huge, looming building was dark. Standing there was a mother, clutching the hand of her little girl. There were no signs of childlike industry, no drawings of families or smiling macaroni faces glued onto paper plates. Instead, there was a feeling of fear and chaos. It was as if no one was in charge. The halls were filled with kids running and yelling. There was a group of kids fighting in the hallway—big kids. It was a shock to Dr. Cantor the first time she saw it, but as she visited more and more schools, she learned that it was typical for schools like this to contain middle-school-age kids who had been held back. They were twelve, thirteen, and fourteen years old, big kids fighting in the hallways right next to kindergarten classrooms. She couldn't help but imagine how the little kids must have felt navigating the hallways of that school every day.

When Dr. Cantor was finally escorted to a classroom, she observed kids making paper airplanes and fooling around, supervised by teachers who looked completely unable to manage their students or control what was going on around them. There seemed to be little or no learning going on.

What the study eventually concluded after many visits to schools across the city and hours of conversations was best illustrated by one of the youngest participants. A five-year-old boy from Harlem was asked to draw a picture of his feelings about 9/11. When he handed it to Dr.

Cantor, she looked first for what she had come to expect: two iconic, smoking towers. They were there in the drawing, but only as two tiny structures in the distance. In the foreground, and much bigger, were two stick-figure children pointing guns at each other.

This picture demonstrated with heartbreaking clarity that to the kids showing the most signs of trauma, 9/11 was only a trigger — two curlicues of smoke on the horizon. The origin of their symptoms was not the acute trauma of 9/11: it was the clear and present danger of their everyday lives; the chronic stress of walking to school through a crime-ridden neighborhood in the morning and then feeling unsafe in school all day meant that the kids in the deepest poverty lived in a state of constant alert.

Dr. Cantor's experience working with children on both sides of town cued her into a critical realization. The communities near Ground Zero were equipped with more resources, which meant adults were far more able to act as effective buffers, keeping the kids' stress out of the toxic zone and into the realm of tolerable. Whether it was a teacher, a religious leader, a grandparent, or a coach, the children closer to Ground Zero had many more sources of buffering that could help stabilize them in moments of acute trauma, even if it was severe.

What Dr. Cantor saw through the research was that poverty itself reduces the resources available to even caring, dedicated parents to be effective buffers for their kids. Not only were children in poverty experiencing a greater incidence of trauma, they were more likely to develop toxic stress because their source of buffering was constrained by the daily existential stresses that families were under. That was what was affecting their ability to thrive and learn in school. And that was the insight that drove Dr. Cantor to leave her practice and dedicate herself to creating solutions that could help very vulnerable children.

When she first set foot in the elementary school in Washington Heights, Dr. Cantor's immediate reaction was burning outrage. As a psychiatrist, she recognized the symptoms of trauma all around her. It wasn't one or two kids, it was *the entire school*. When people hear the word *trauma*, they often think it represents a small percentage of children requiring services in a typical high-need school setting, somewhere around 10 to 15 percent of the kids. That was what Dr. Cantor once believed. What she came to learn after visiting many high-need schools was that while there might be a relatively small percentage of kids who needed individualized mental-health services, the students who required *something* beyond a traditional educational environment in order to be able to be ready and engaged in learning was much, much larger.

Turnaround for Children was founded after 9/11 with the recognition that while most schools inherently acknowledge the importance of mobilizing resources in response to *acute* traumas, they simply aren't set up to address the insidious ways in which the day-to-day onslaught of chronic adversity undermines learning. First the organization had to educate people about the connection between adversity and academic performance. Despite all the research, Dr. Cantor and her team still found that this wasn't always intuitive for many educators. Next, Turnaround had to figure out how to support schools in designing practices and interventions that worked for kids dealing with the impact of stress to improve their learning outcomes. No easy task.

As a physician, Dr. Cantor approached the problem via the neurobiology of adversity. In order to be able to pay attention and learn in school, a kid needed to engage his prefrontal cortex (the conductor), which meant the amygdala alarm had to be silent. Safety and stability would be key components to the solution. But how could Turnaround create safety and stability in the classroom when kids were bringing these stressful experiences from home and the community into the classroom with them, causing problems and challenges for teachers and fellow students? Dr. Cantor and her team knew that for many of the kids they were serving, the amygdala alarm was always on high alert, and the cortisol thermostat was overheating. They also knew that the natural antidote to toxic stress — having a well-regulated caregiver who could buffer the stress response — was often in very short supply.

Turnaround began by using the science to inform school practices and policies. They placed mental-health professionals and social workers in schools, building systems of support that families could easily plug into. Turnaround invested in training every adult in the school environment, from the leadership to the guidance staff to every single teacher — because they recognized that the traumatic effects of adversity crossed an entire school building. They observed that one child in a classroom with attentional and behavioral challenges will often disrupt a lesson, but *thirty* children with these kinds of struggles can trigger a tinderbox effect, shutting down learning for everyone.

One of the biggest challenges for many schools was discipline, how to balance the safety of the school community with the needs of each individual child. The traditional model of school discipline was reactive and punitive (you do X, the consequence is suspension or expulsion), and that meant that a lot of kids were losing valuable time in the classroom. Turnaround developed strategies aimed to work *with* a student's biology instead of against it by first addressing the dysregulated stress response and *then* dealing with the issue at hand. This could be something as simple as offering a student a better choice for dealing with a stressful moment, such as retreating to a space reserved for quiet reflection or prompting a student with a silent signal to count to ten and breathe deeply.

Their approach had a profound impact on school culture. Across Turnaround partner schools from 2011 to 2014, suspensions were cut in half. Measurements of classroom climate, productivity, and engagement jumped by over 20 percent, and severe incidents declined by 42 percent. Dr. Cantor and her team expanded Turnaround to more cities, bringing their model from New York City to Washington, DC, and then to Newark.

Still, they found themselves struggling with one especially frustrating challenge. All of the science suggested that the positive outcomes they were seeing should pave the way for improved learning, but despite all the wins on school culture and climate, test scores remained surprisingly stubborn. They racked their brains for what they could be missing. They met with school leaders, looked at their data, and went to educational conferences to learn from others' best practices.

The breakthrough from Dr. Cantor's perspective ultimately came with a shift in *how* they looked at solutions. She saw that educators often lifted up one practice as *the* solution to the problem. After being in the education world fifteen years, Dr. Cantor had seen how accountability and measurement was now *the* thing, how expectation was *the* thing, how a great teacher in every classroom was *the* thing.

It hit her that in medicine, she hadn't been trained to ask, What is *the* thing? Her training told her to ask herself: *What explains the symptoms we're seeing?* And usually the answer was more complicated than just one thing. She realized that Turnaround had to apply interventions based on a comprehensive understanding of the problem. It was tremendously important for kids to go to a school where they felt physically and emotionally safe. *Check.* It was also really important for kids to develop their readiness for learning, because the exposure to adversity affected the skills that were involved in learning readiness. *Check, we have to do that too.*

Many school systems were profoundly influenced by the realization that when it comes to student success, teaching things like resilience and grit can be as important as teaching math and science. Dr. Cantor and her team went one step further. The developmental neuroscience suggested that before kids could learn grit and resilience, or math and science, for that matter, they needed a basic foundation in healthy attachment, stress management, and self-regulation. Healthy attachment is what Dr. Lieberman and Dr. Renschler worked so hard on with Charlene and Nia. When it goes right, healthy attachment begins at birth and forms the basis from which we all learn to trust and relate to one another. For many children, growing up in poverty, in families stressed by economic and other insecurities, healthy attachment and stable nurturing experiences were much more challenging. Whether it was chaos at home, violence in the community, the crushing weight of poverty, or the fog of drugs, alcohol, and mental illness, families often faced overwhelming challenges in providing safety and security for their children.

Dr. Cantor realized that they had created a model built on a foundation that many of their students never got, which was why their model was only partially effective. They figured out that when it came to educational success, the key was not just to provide the right ingredients; just like with Tyrone's tadpoles, the *timing, sequencing,* and *dosage* of these ingredients was critical.

So, Turnaround came up with a framework it called Building Blocks for Learning that worked to develop in children the foundational skills of attachment, stress management, and self-regulation, and *then* layered the other skills for learning on top. By ensuring the development of these skills in an order that makes sense for learners' biology, Turnaround was building on decades of neuroscience telling us that it's not enough to "step on the gas" by providing enriched environments to support learning for children. You also have to release the "brake" (the inhibitory effect of the amygdala on cognitive function) by supporting attachment, stress management, and self-regulation. In doing so, Turnaround may finally be able to crack the notoriously difficult test-scores problem for kids living with adversity. Their partner schools in the Bronx are beginning to see net gains in scores in math and language arts that outpace the gains of other schools in the district.

Far from stigmatizing and singling out kids with ACEs, Turnaround embraces an approach that simply identifies where a student is on the developmental trajectory and uses the science of toxic stress to help get that child back on track. Knowing whether a kid's development is stuck because of exposure to ACEs is fundamental to figuring out where to start in the classroom.

Dr. Cantor's description of her schools was consistent with everything I knew about toxic stress. I thought about my kids in Bayview, the ones whose learning and behavior problems in their classes were so often severe. It hit me that ACEs weren't just at the root of a public-health crisis in America, they were at the root of our *public-education* crisis as well.

It was clear that while ACEs might be a health crisis with a medical problem at its root, its effects ripple out far beyond our biology. Toxic stress affects how we learn, how we parent, how we react at home and at work, and what we create in our communities. It affects our children, our earning potential, and the very ideas we have about what we're capable of. What starts out in the wiring of one brain cell to another ultimately affects all of the cells of our society, from our families to our schools to our workplaces to our jails.

Nancy Mannix, Jeannette Pai-Espinosa, and Pam Cantor were taking this new understanding and integrating it into their work in ways that were creating breakthroughs for the communities that they served. Despite the pushback and the naysaying, these women were on the vanguard of the movement, slowly but surely bringing ACE-informed approaches to scale.

I made a mental note to stay in touch with these women, to learn from their successes (and failures), and support and encourage them in any way that I could. I felt heartened to see the movement gaining traction beyond the field of pediatrics and branching out the beginnings of a true public-health movement. Still, I felt unsettled. It was unnerving how quickly the conversation at the conference had gone sideways. I knew that what I really needed to understand about that conference was *Why the hateration?*

. . .

A few weeks later, I found myself once again packing up my breast pump to attend yet another conference that I just couldn't miss. This one, hosted by the White House and the Gates Foundation, was being held at the University of California, San Francisco, which meant that at least I didn't have very far to travel. As I handed Grayboo to my husband with a kiss and then stepped out the door, I found myself looking forward to this conference more than I had for almost any other in recent memory. I wasn't speaking, which felt a little bit like a luxury. I could just sit back and soak up all the exciting new research and delicious data.

The agenda of the Precision Public Health Summit was to bring everyone together to discuss how precision medicine could be used in the public-health arena to level the playing field in the critical first one thousand days of a child's life. In other words, it was right up my alley. The discussion was wide-ranging, but a big theme throughout was the importance of partnerships between scientists and the communities they are trying to help. One of the speakers from the community partner side of things was Jenee Johnson, the director of the Black Infant Health Program (BIH) in San Francisco.

The organization's mission is to improve maternal and infant health in African American communities, which meant our paths had naturally crossed. Even before the Bayview clinic opened, Jenee recruited me to lead a class on common health concerns for babies that BIH hosted at the Bayview YMCA. All these years later, I was happy to see BIH's wonderful work represented at the summit.

But soon, as someone who is pretty conversant in the worlds of both science and community, I noticed a natural tension playing out in front of me. The researchers and statisticians who sat beside Jenee talked about biomarkers and data sets, about the difficulties of data collection and privacy. Jenee, however, spoke passionately about the moms and babies she worked with and the day-to-day reality of poverty and social adversity in the community. She talked about respect for black women, clapping her hands as she repeated "Respect, respect, respect," emphasizing each syllable and raising her voice with each clap. To the research scientist, numbers are people. To a person who serves vulnerable families, numbers distract from real experience.

As she began to speak to the audience, the emotion in her voice made the room of over three hundred scientists feel very, very small. Jenee talked about a mother who showed up to a program one evening with all of her possessions in a suitcase and a baby on her hip because she had nowhere to spend the night. Her voice rose with pain and anger as she talked about how science was failing the people she worked with by not putting them at the center of the work.

"What's the serum for helping a community to stay together and not be dismantled? I have families that now commute back to my program from Antioch — forty-five miles away. What's the serum for that? Dr. Martin Luther King told us that it does not cost America *anything* to have me drink at the same water fountain as you. It does not cost America anything to have me sit at the front of the bus. But it is going to cost *something* to make sure that we have educational equality, equity in jobs, housing. So we are gathered here, and this is a beautiful gathering, but we are missing a whole other group of people. Because to manage stress, the stress that my clients come into the office with, I don't have a serum, there's no pill, there's no research question to help me help them. We keep talking about *stress, stress, stress,* and let's *study, study, study,* when the axiology of black people is *relationship.* We all know that. We need to bring them up on the agenda and bring other people into the space. Especially the people that this impacts. We're at meeting number five hundred that I've been to, brother, and they are not here."

The room remained silent for a moment, and in that small slice of time, a surge of conflicting emotions overcame me. I felt Jenee's anger about the lack of diversity in the conversation and her heartbreak for the young mother who had nowhere to go. I agreed with much of what she had said, but her statement that the people affected by stress weren't here was dead wrong. I knew that for a fact. For a split second, my husband's face flashed in my mind. His expression was taut with alarm, his jaw clenched — he looked menacing in a way I'd never seen him before.

• • •

It was 2014, before Grayboo was born, and we were at Lake Tahoe in Nevada with the kids waiting for a table at a restaurant. I remember rounding the corner returning from the restroom and catching a glimpse of my husband. His appearance was alarming. I took in every detail of the scene as if it were playing out in slow motion. His body was tight as a drawn bow, full of potential energy that looked like it was about to become kinetic. His fists clenched and unclenched. I could see fat wormlike veins standing out on his forearms. His eyes, shifting back and forth, were trained on our three rowdy black boys playing in their usual oblivious manner on the bench in front of the restaurant. Kingston, only two years old at the time, was trying to push my twin stepsons, Petros and Paulos, both eleven, off the bench. He was laughing and shoving, and they were doing their best to goad him into showing his fiercest feats of strength. Then Arno's eyes led me to look just past them, to two burly Caucasian men with shaved heads, steel-toe boots, and dark gray-blue tattoos snaking up their necks. The men were glowering at our sons. I recognized immediately that Arno was in full fight-or-flight mode, and for a second, I thought my own heart might stop.

Just then, the hostess called our name, giving us a good reason to get away from the two human bears in the forest. But the image of my husband in that moment, bare-knuckled and ready to brawl as he watched the men glaring at his kids, is burned in my mind for two reasons. One is that as the father of black children, Arno has an additional risk factor for stress. When you are black or brown and living in America, there are more threats and stressors inherent in your experience; in other words, you live in a part of the forest where there are a lot more bears. Race is never easy to talk about, but exposure is exposure is exposure — that was a big part of what Jenee had been saying, and she was right.

But the other reason I will never forget that moment in Tahoe is the thing I wished I could share with Jenee: While he does have black kids, my husband is white. In fact, my sweetheart, Whitey McWhiterson, the Mayor of Caucasia (as I affectionately call him), is both white *and* a successful CEO. He sits on the top of the socioeconomic food chain. If you were to look up *the Man* in the dictionary, you'd see a picture of my husband. My two stepsons are adopted; their complexion is darker than mine, while Kingston is a creamy caramel. Undoubtedly, the two men snarling at our kids had no idea that they were standing just a few feet from their father. But in that moment, Arno was just a dad whose kids were being threatened. What I saw was a profound example of the intersection of biology and society. The stress-response mechanism is hardwired into all of us. Threat equals reaction, and it doesn't matter if the threat is in the form of a Confederate flag tattoo or a strapping grizzly; the same biological mechanism is triggered.

What I felt Jenee wasn't seeing was that while my kids and hers might have stress-response-triggering experiences because of their race, poor white kids living in Appalachia also have triggering experiences. Think about it like this: We all live in a forest with different kinds of bears. There is a large group of bears that populate a part of the forest called Poverty, and if you live there, you're going to see a whole lot of bears. There's also a part of the forest called Race, where a different cluster of bears hang out. And there is another bear neighborhood called Violence. If you live near any of these bear dens, your stress-response system is going to be affected. But here's the important part — it is affected in the *same way* no matter which bear you tango with. Unfortunately, a lot of people (like my patients) live in a place in the forest where the neighborhoods of Poverty, Race, and Violence overlap, and for them, it's wall-to-wall-to-wall bears. But there are also a lot of bears that live in the neighborhoods of Parental Mental Illness and Divorce and Addiction, which is why I reacted so strongly to the last part of Jenee's statement. Some of "the people that this impacts" *were* in the room.

That's why we need to collect broad swaths of data, because public-health-scale solutions require us to identify and measure toxic stress in everyone, not just one group of people. We are not going to make a dent in this problem by creating solutions for just one community.

Suddenly, as I sat listening to Jenee, something in me shifted. It was as if someone had flipped a switch. This was it! This was precisely the root of so much of the emotional blockage around ACEs that I had encountered. It was why those folks in New York got so riled up so quickly about the thought of *their* kids being stigmatized by screening. And right now, the anxiety and pain was etched on Jenee's face. *What about us?* she seemed to be saying. *What does all this do for the pain and suffering in* my *community?* That sentiment is both totally understandable (the pain and suffering of the African American community is one of our country's deepest unhealed wounds) and exactly what will keep us running in place for years to come.

I stood up, trembling.

With the hush in the room, I didn't need a microphone.

As I talked, I could hear my voice shaking. I may have been speaking to Jenee and the others in the room in that moment, but it felt more like I was screaming on the edge of a canyon, hoping the echo would carry for miles.

"I think that all of us are in this room because we are trying to come up with the solutions for the *entire* population. Some of it has to do with payment for mental-health services so that the parents of my patients who have mental-health disorders can get good enough care so that they can hold down a job so that they can keep their children in housing. I believe that when we make the connection between adversity and only the people who you are seeing and I am seeing every day, our stories are not enough. We need to connect our stories with the science and the data."

My voice rose. I could hear my *T*s becoming more crisp, my *A*s opening, and my *can*s turning into *cyan*s as the lilt of my childhood patois dialect undercut my attempt at composure. Tears welled up and spilled down my face.

"It's not just dat it doesn't cost Americah anyt'ing for us to drink at di same watah founten. We mus' *show* dat it costs Americah *bilyons* of dollahs in cardiovasculah disease and cyan-sah and housing and educya-shon for us to drink at diff'rent watah fountens!" The room erupted in applause.

"We need to make dat argument! We mus' hexplain to ev'ry person dat if dey are in Appalachia, if dey are living in Middle Americah, if dey are living in Kentucky and dey believe dat dey have it hard — we mus' mek sure dat ev'ry single person knows dat dey cyan get strong solu-shons — for poor white folks and for de peer-ent who brought her child and her syuitcases to you — dat we are in a *united* struggle about de effects of adversity on de developing brains and bodies of children. And when we *all* get behind dat, *den* we will have solu-shons that will lif' ev'rybody up!"

I sat down, trembling with emotion. When Dr. Clarke handed me Dr. Felitti's paper almost ten years earlier, I had been able to pull the pieces together and recognize what was really going on with my patients. In that moment at UCSF, my heart still racing, I realized that I had just had a second (very public) epiphany. Why were people so resistant to the science of adversity and to giving a basic fact of our biology a name and a number? Because when you bring it down to the level of cells, the level of biological mechanisms, then it is about *all* of us. We are all equally susceptible and equally in need of help when adversity strikes. And that is what a lot of folks *don't* want to hear. Some want to stand back and pretend that this is just a poor-person problem. Others take fierce ownership of the problem and say, "This is killing my community," but what they also mean is *It's killing my people more than yours.*

In rural white communities, the story is about loss of living-wage work and the fallout from rampant drug use. In immigrant communities, it is about discrimination and the fear of forever being separated from loved ones at a moment's notice. In African American communities, it's about the legacy of centuries of inhuman treatment that persist to this day — it's about boys being at risk when they are playing on a bench or walking home from the store wearing a hoodie. In Native American communities, it is about the obliteration of land and culture and the legacy of dislocation. But everyone is really saying the same thing: *I am suffering.*

It is easy to get stuck on your own suffering because, naturally, it is what affects you most, but that's exactly the mentality that is killing black people, white people, and all people. It perpetuates the problem by framing it in terms of *us* versus *them.* Either *we* get ahead or *they* get ahead. That leads quickly to a fight for resources that fragments efforts to solve the same damn problem.

What I was trying to communicate to Jenee and to everyone in the room was that this very human instinct toward tribalism was why we needed science. That was why we needed every researcher, data cruncher, and scientist in that room. Because the science shows us that it is *not us* against *them.* In fact, we all share a common enemy, and that common enemy is childhood adversity. The approach to treatment for the homeless child standing with her mom holding their bags at a Black Infant Health Program meeting is the same approach you use for the family in Pennsylvania where the dad hasn't worked in five years because the plant closed down and for the little girl in rural China whose mother had to leave her to find work in Beijing and for the families in Montenegro and Serbia who lived through civil war . . . It's the *same* fundamental approach to treatment for us *all.* If we begin to understand that, then maybe we will stop being so Balkanized in our response to the problem and be able to come up with solutions that work for everyone. Because, as my dad used to say in his Jamaican patois, "That rising tide, she lif' up *all* di boats, mon."

12

Listerine

IT WAS 1:00 P.M. on the nose when I strode into the clinic, the last few bites of my lunch in a brown, compostable to-go box. I thought I had a few minutes to spare before my first patient of the afternoon, but as I passed the reception desk, Nurse Mark flagged me down.

"Your first patient is roomed and ready," he said, handing me a printout of my notes from the previous visit as well as all the new paperwork my patient had completed for today's appointment. "They were early, so I put them in the butterfly room."

"Got it," I replied, scampering off to the physicians' office to quickly throw on my white coat and grab my stethoscope.

I couldn't help but smile to myself. It had been ten years since we opened the Bayview Child Health Center. I couldn't have imagined in 2007 that in 2017 we would still be in Bayview . . . that I would still be here. I certainly wouldn't have dreamed that the Bayview clinic would have inspired the creation of the Center for Youth Wellness or that the two organizations would be working side by side not only to screen every child for ACEs and provide comprehensive care but also to share our tools, models, and clinical insights with doctors around the world. As things evolved, the one constant was the dedicated and caring nature of our staff. When Nurse Mark came on, he took over the day-to-day management of the clinic, which meant that although I was the founding physician, he ran the show. I was taking my orders from him.

A few minutes later, after my standard "Knock, knock," I stepped through the door of the butterfly room and into what remained, by far, my favorite part of the week — seeing patients. The butterfly room is named for the hundreds of little butterfly decals all over the walls, thoughtfully placed to make it feel as if they are all flying toward some beautiful invisible flowers down the hall. When the Bayview clinic moved into the Center for Youth Wellness building in 2013, the staff had taken pains to ensure that our new space was as welcoming and kid-friendly as the old one. Every room was decorated with dozens of wall decals, each in a different animal theme; there was the jungle room, the dinosaur room, the safari room, the under-the-sea room, and the farmville room. But the butterfly room was definitely my fave. When I first saw it, it had taken my breath away. Most of the butterflies were flat stickers on the wall, but a few, up in the corner above the sink, were in 3-D, their pink and purple wings protruding as if to say, *We're real!*

My sixteen-year-old patient was perched on the exam table, his eyes glued to his phone. He was busy whipping out a text message or Instagramming or Snapchatting or whatever sixteen-year-olds are doing with their phones these days. His mother was sitting in the chair next to the sink clutching a small piece of paper with some handwritten notes.

"Hey, guys! How are you?"

My patient looked up and gave me the same sweet smile I had known for almost ten years. Well on his way to manhood, he was slim and muscular with a barely there line of fuzz on his upper lip. Always neatly put together, he wore freshly ironed khaki pants and a tucked-in white shirt. He'd let his usual crewcut grow out a little in the front, and I noticed a skillful upward sweep of pomaded bangs that I'd never seen before.

His response was the typical one-word utterance that is the hallmark of teenage communication: "Hey."

I smiled to myself, mentally filling in the first of many boxes: *Language developmentally appropriate? Check!*

I sat down on the little wheeled stool in front of the computer and reviewed his latest information. By then, I felt like I knew his chart by heart. It contained an ACE score of seven, symptoms of toxic stress, a history of successful interventions over the years when things flared up, and all of his most recent labs. The last time he came in, about a year ago, he had been in a great spot physically and mentally. His asthma and eczema were under control and he'd been doing well in school; he was even developing his first fledgling relationship with a girl. The toothy grins and laughter of childhood had given way to reluctant (though still boyish) smiles and a baritone voice. I could almost see the hormones coursing through his body.

Though he had flashed me a reflexive smile when I first walked in, I knew when I looked at his mother that there was a concern. The expression on her face was as valuable as anything in the medical chart that day. Her brow was knit with the same mix of worry and hopefulness that I had come to know over the years. Something was going on.

Fortunately, by now, Diego knew the drill.

It was time for a tune-up.

Like many of our patients with ACEs, he had gone through an initial period of intense therapy and other medical interventions when he first came to see us. We had successfully managed his asthma and eczema, and he'd resumed a normal growth velocity, though he never completely made up for the years of growth he'd missed. As his primary medical home, we were there for every bump and scrape moving forward. Because the impacts of early adversity are chronic and long-lasting, rough patches are inevitable. That's something Diego and his mom had learned to take in stride; they understood that his stress-response system was going to need a little TLC from time to time. Helping him navigate the medical services and therapies he needed was where I came in as his doctor.

When I asked Diego what was up, he knew what I really meant was *If something is triggering your stress response, we've got to get on it early. Is there anything going on that we can help with?*

Taking a deep breath, little Diego, who wasn't quite so little anymore, looked up at me.

"Um, I don't know," he mumbled, and then he looked at his mom.

"Doctora," Rosa began, smoothing out the crumpled piece of paper, "*necesita su ayuda.* He seems depressed. He's missing classes. His grades have gone to Ds and Fs. I know my son is struggling. He needs help."

I looked at Diego. "Is that right?"

He nodded sheepishly.

I asked Rosa to step into the waiting room and then I slid my stool over to Diego and rested my hand on the edge of the exam table.

"You want to tell me what's going on?"

It turned out the girl he had been seeing for the past year had problems of her own. She had been going through some family stuff and it was taking a toll on their relationship. With her, it seemed like everything was up or down, on or off. Either their relationship was awesome, the best thing in her life, the thing that was saving her from everything else, or it was awful and there was no way it could work out. A short time into their relationship, Diego found out that she was cutting herself. She didn't want him to make a big deal out of it. It was just something she did when things felt like they were too much. But Diego couldn't bear it. He wanted to protect her, from her family and from herself. He wanted to be the unconditionally accepting caregiver that her real family wasn't. So he started going over to her house every day after school to hang out, but it was a rough place to be. Diego didn't want to be there, but he couldn't leave her alone either. Soon, the yelling and the drama took him back to a familiar, dark place.

Even before adolescence, Diego had gone through periods of suicidality. One night when he was eight, his dad got drunk and attacked his mom. Fearing for his mother, Diego called 911. The police came and arrested his dad. Because he was undocumented, his father was soon deported to Mexico.

Diego felt horribly guilty for calling the police on his dad. He had just been trying to protect his mom, but now his dad was gone. The very thing they had always feared. Everything became harder. His mom took on another job to make ends meet, but it wasn't enough. Diego and his mom and little sister moved into a smaller apartment to save money, but they still went hungry sometimes. Diego missed his dad terribly and stayed in close touch, writing him regularly and calling when he could. In every letter and on every phone call the question was the same: *When are you coming home?*

Then his dad's letters stopped coming. The phone went silent. Weeks went by — nothing. Diego feared that his dad might be angry with him for calling the police. He wondered if his dad had gotten a new family in Mexico and didn't care about him anymore. He asked his mom if she knew where his dad might be, but his questions just seemed to make her sad, and she didn't have any answers. Finally, months later, Rosa heard from one of her cousins. Diego's dad was a *desaparecido* — one of the many who vanished after resisting the Mexican drug cartels.

Shortly after hearing this news, Rosa got a call at work from San Francisco's child-crisis response team. Diego had somehow gotten himself onto the roof of his school building and was standing near the edge, crying so hard his whole body shook, saying he didn't want to live anymore. He was there for over an hour, sobbing as he stood less than a foot from the edge of the roof. Finally the child-crisis worker coaxed him into her arms and took him back down to safety.

His mother quickly brought him to the clinic, and we were able to plug him back into therapy with the same clinician he'd seen before, someone he knew and trusted. These dark periods for Diego were hard to deal with, but as time went on, he learned to build more and more strategies to help mitigate his symptoms when rough patches cropped up. As his doctors, we found that the ACE lens made it relatively easy for everyone on his medical-care team to coordinate with his mental-health and wellness teams.

So a few years later, when Diego was twelve and came in with the worst asthma attack he had ever had, our multidisciplinary team was there to help. Struggling to make ends meet, Rosa had moved the family into a rundown old apartment. Though it wasn't much, it kept them close to the friends, schools, and medical care that were helping her kids stay on track. One night an electrical fire started in the kitchen. When I heard about the fire, I assumed that Diego's asthma flare-up was a result of the exposure to the smoke from the blaze. But days later, at his follow-up appointment, he was still having severe symptoms despite strong medications, and I knew that I needed to ask if there was more to the story. It turned out that Rosa had gotten Diego and the kids out of the apartment very quickly. Diego had been exposed to almost no smoke at all. But the fire that claimed their apartment had left them homeless, and he and his family hadn't eaten for three days. Diego took it upon himself to be the man of the family, to protect his mom and his little sister and provide for them. But at twelve years old, he was terrified. No matter how strong he wanted to be for his family, the nights on the street were taking their toll on his biology. It was only after our social worker found emergency housing for the family that I was ultimately able to back off of the high-dose asthma medication.

So when Diego told me about his girlfriend and her family, my heart ached over this new episode of sadness and pain in his life, but I was also confident that we could help him through it. By then we had a sense of what seemed to work best for him. Rosa knew what to look out for in terms of changes in her son, and Diego knew that when he was feeling really bad, our team would stick with him until he felt better again. As he always did, Diego gave me a hug when he left the exam room that day, and this time I made sure to give him an extra squeeze back.

Over the next weeks, our team worked with Diego to assess how he was doing in the six critical areas of sleep, exercise, nutrition, mindfulness, mental health, and healthy relationships. We knew Diego would benefit most from an intense regimen that involved seeing his long-time therapist at CYW, and we also identified someone at his school he could check in with on a regular basis. I encouraged him to rejoin the pickup soccer games he loved and connect more with his sources of support, including his mom. It wasn't long before we started seeing improvement. Ultimately, the relationship between Diego and his girlfriend ended, and over time his grades crept back up to As and Bs. He even made the honor roll. He decided that he might want to be a lawyer and landed an internship at the district attorney's office, which he totally loved. Then there was the new puppy that he'd adopted, the one he glowed about when he told me of her antics. He loved taking care of her, and when he scratched her ears, she licked his face.

Months later, when I saw him for a follow-up appointment, I felt a deep sense of satisfaction at the progress he had made. The system was working exactly as it should. Diego was back on track.

If this were a movie, we could roll the credits right here and all of us could feel pretty good about ourselves. Diego had "made it."

But that's not how life works. The story doesn't stop.

In real life, Diego lives in a dangerous neighborhood, and things keep happening.

A couple of months later, I saw Diego's little sister in clinic for a checkup. Still in diapers when I first started to treat her and Diego, she was now almost eleven years old. Rosa came with her daughter, and on their way out the door I asked her how Diego was doing.

By then I had come to know Rosa's sighs pretty well. There was the one with an extra-long exhale that meant she was exhausted, and then there was the short, exasperated one that she let out when she felt frustrated or confused. The one she let out that day was deep, and she closed her eyes as she exhaled and put a hand to her chest. The quality of it reminded me of the first day that we had met, before she told me then-seven-year-old Diego's story.

"*¡Ay, Doctora!*" she said. "I know my son very well, Doctora. I watch every detail of him. I know how he reacts to everything. I'm like a detective, watching him, but not too close. It's not easy."

"Did something happen?" I asked.

"About two weeks ago, I knew something was wrong. I could see he was about to fall into depression, so I started asking him, *Are you okay?* He just says, *Yes, Mama.* But I keep watching what he does and I know something is not right, so I say, *Mi amor, I see you. You are just sleeping, you don't want to bathe yourself, you are not eating. I can see that you are suffering. Tell me, has something happened?* But he says, *No, Mama, I'm okay.* It was Saturday afternoon, and I had to go to Mass, so I said, *Why don't you come with me?* He says, *No, Mama, I want to stay.* I was about to cancel going to Mass. I was not calm because I knew that my son was going through something, so I go into his room and I say, *Son, are you depressed? I will stay here with you.* But he told me, *No, Mama, estoy bien, you can go.* So I went. During Mass he sent me a text message saying, *I'm sorry.* I could not understand the rest because it was in English, so I showed it to one of my friends next to me who speaks good English and asked her to read it to me. It said, *Mama, forgive me for what I am about to do.* Doctora, let me tell you, I was sitting in that church in Oakland, and I felt *such* an anxiety. In that moment, if I had a magic wand, I would disappear and appear in my house. I went into a panic. I imagined getting home in forty-five minutes and finding my son dead. I had to find a way to return to San Francisco. I begged my friend who has a car to drive me home right away. Those were the most stressful minutes."

Rosa's voice choked. Her eyes brimmed with tears.

"I called him, but he didn't answer me. I sent him texts and still he did not answer. I even borrowed my girlfriend's phone so that I could call him from a number that wasn't mine, but he didn't answer that either. His phone rang and rang but he didn't answer.

"I have a friend, Magdalena, who lives nearby. I wasn't expecting her to be home because she usually is out dancing with her boyfriend on Saturday, but *gracias a Dios,* she was home. I told her to go to my house, that the life of my son was in her hands and I needed her to help me save him. I begged her, I said, *Magdalena, go to my door and knock on the door and keep knocking until he answers.* She knows that he suffers from depression so I told her to call the police if he doesn't answer.

"Doctora, in my neighborhood, we do not call the police on each other, but I told her, I begged her, if he doesn't answer, you must call the police. She told me not to worry, that she would do it. Every moment that passed I felt like he was slipping through my fingers. I was crying with anguish. I called him again and again.

"Finally, when we were about halfway home, he answered the phone and I asked him, *Mi amor, are you okay?* But he didn't want to talk to me, so my friend took the phone and asked him if he was okay. She told him, *Your mother is worried! She doesn't deserve to suffer, answer her! ¡Contestala!* But he was silent. He listened but didn't say anything, so she told him that Magdalena was coming and that the police would break down the door if he didn't answer. When I got home I was trembling. When I found him, he was on the floor. I thought that maybe he had taken pills. *Gracias a Dios,* he didn't. He was *en una buena borrachera* — he was really drunk. That was all! He had a bottle of Bacardi Silver and was very drunk and feeling very bad. That's how I found out his friend had died."

"Oh my goodness!" I gasped.

"*¡Sí, Doctora!* A good friend of Diego. He had just graduated and was walking on the street with another friend when someone shot him. He was a nice boy. A good student. He never got into any trouble. The bullet was not meant for him, but he was the one who died."

"I'm so sorry," I said.

"*Gracias*. But Diego is okay. I made him call his therapist that same day, and she is helping him. He's doing better now, but Doctora, I tell you, it's not easy."

. . .

I followed up with Diego's therapist later that day to be sure that he was getting the help he needed, but I felt sad and angry and frustrated all at the same time. Only months before, he had gone through the rough patch with his girlfriend and come out on top. The last time I spoke with him, we had joked around about his internship at the DA's office and I had asked him where he wanted to go to college. And then, in an instant, a kid just like him, someone he knows and cares about, is walking down the street and is in the wrong place at the wrong time and is gone forever.

I get a bad feeling in the pit of my stomach knowing that this will undoubtedly happen again. Not the same incident, but certainly one with a similar impact. Some triggering event will expose Diego to a level of stress that trips his already sensitive stress-response system. Even with all the progress he's made, it will likely send him flying. He will have to try to keep his head about him to the degree that he can, recognize what's going on biologically, and marshal his resources. For now, he has his mom to help him do that, and she has the clinic to help both of them. And that's the good news. That's the reason we created CYW in the first place. It's what we can do. We can't erase Diego's past trauma or build him a protective bubble to float through life in, but we can use what we know about his biology to mitigate the impacts of the toxic stress that will forever be a part of his world.

We were giving Diego state-of-the-art care. The problem is that the state of the art sucks. Compared to what we *know* about the mechanism of toxic stress, what we *do* is still rather primitive. I wished we had better diagnostic tests to figure out exactly which pathways were being most disrupted so we could target our treatments more effectively. I wished we could wash the impacts of toxic stress from his DNA the way that Michael Meaney had done for his adult rats — wash away the imprint of adversity, wash away the risk of asthma and suicide and heart disease and cancer.

I thought about my days at Stanford on the pediatric oncology ward. I wished we could do for Diego what we had done for my patients with leukemia. At Stanford, when we treated a patient with cancer, everything was done by protocol. POG Protocol #9906 was for high-risk acute lymphoblastic leukemia that had spread to the central nervous system. If the brain and spinal cord weren't involved and the cancer was less aggressive (a white blood count of less than 50,000), then POG Protocol #9201 could be used. The three letters *POG* before each protocol number was something that I didn't think about much at the time. It wasn't until Diego and others like him sent me on my quest to understand and treat toxic stress that I stopped to wonder, *How the heck did they know which interventions to use?*

In 1958 the survival rate for childhood cancers was 10 percent; 90 percent of kids diagnosed with cancer died. By 2008 the survival rate had been raised to almost 80 percent. Patients with acute lymphoblastic leukemia went from a six-month median survival (meaning only half of the patients survived six months from diagnosis) to an 85 percent overall cure rate. How on our Lord's green earth did we as a society manage to do that?

Well, as it happens, the answer to the question lies in the three letters before each protocol number. *POG* stands for Pediatric Oncology Group. It was one of four pediatric clinical-trial groups dedicated to treating childhood cancers; all the groups merged in 2000 to create what is now the Children's Oncology Group (COG). Today, COG membership includes over five thousand pediatric cancer specialists in approximately 230 medical centers in the United States, Canada, Switzerland, the Netherlands, Australia, and New Zealand. At COG institutions, multidisciplinary teams consisting of physicians, basic scientists, nurses, psychologists, pharmacists, and other specialists use their skills in the investigation, diagnosis, and management of childhood cancer.

This groundbreaking collaboration resulted in the development of a successful multidisciplinary model for care, more effective cancer therapies, and carefully refined care protocols that help patients get better, faster. It wasn't one or two labs doing cutting-edge research that tipped the scales. It wasn't the development of a single pill that made the difference. It was the spirit and practice of collaboration across the United States and, indeed, the world. The cancer specialists shared a goal, but just as important, given how intensely competitive and resource-constrained academic medicine can be, they shared patient data, ideas, and research.

But researchers didn't collaborate just because they were moved by the spirit to cure pediatric cancers (although I am sure that they were moved). In 1955, the National Cancer Institute (NCI) decided that the study of leukemia could move forward more quickly if researchers came together in "cooperative groups." The organization modeled the program after a successful effort by the Veterans Administration that incentivized collaboration among researchers working on advancing care for tuberculosis. In 1955, Congress allocated five million dollars for the NCI, which ultimately led to the creation of seventeen research collaboratives that transformed clinical practice and dramatically improved outcomes for pediatric cancer patients. By the time I was a resident on the peds oncology ward at Stanford, I could assure parents that although childhood leukemia was a very scary diagnosis, the disease was eminently treatable.

When you compare toxic stress to pediatric cancer, its treatment is still nascent — we are just beginning our response. If the global crisis of childhood adversity were a book, we would be in the second chapter. In a lot of ways, *this* book is the story of that first chapter — the discovery of the biological mechanisms. We have not yet perfected our response. But we're working on it. CYW just took its first baby steps toward developing the kind of research partnerships that lead to breakthroughs in patient care. Working with some heavyweight research institutions, our teams are doing the type of rigorous randomized controlled trials that are necessary to answer big questions like "Can we find biological markers for toxic stress that can be reliably measured?"

How do we move from having the first piece of the puzzle — knowing that adversity leads to a damaged stress response, which leads to toxic stress, which itself is the driver of a whole host of negative biological impacts and disease states — to the type of public-health enlightenment that I had read about in grad school? For me, this frame shift is on the same scale as the medical community's acceptance of germ theory, and in fact, medical history offers a compelling road map for the future.

. . .

Back in the day, before medicine recognized that infection was caused by microbes, people thought it was caused by foul air. While this may seem ridiculous to us now, in nineteenth-century England, it was supported by the observation that the more chamber pots that were dumped out into the street every morning, the more likely there was to be an epidemic of cholera. Similarly, when surgeons approached a patient with a severely infected wound, the smell test was an important piece of diagnostic information. The more putrid the wound smelled, the more likely the patient was to die. Scientists of the day hotly debated the causes of epidemics like cholera and the Black Death (bubonic plague), but the most widely held belief was the miasma theory of disease, which postulated that poisonous vapors arose from rotting matter and made people sick.

Until the late nineteenth century (and, actually, into the early twentieth century), clinicians and scientists believed that the best way to prevent infection was to get rid of the bad smells. And they were partially right, so their treatment was partially effective. Minimizing the dumping of raw sewage into the streets and water supplies did reduce the risk of cholera. But placing flowers in the surgical masks of doctors and by the bedsides of sick patients did nothing to reduce their risk of death (though the latter is a practice that we still follow to this day).

One big problem with the miasma theory was that if something didn't smell particularly bad, folks figured that it couldn't be the source of disease. This was the case with the well on Broad Street investigated by Dr. John Snow. Because the well water didn't smell awful, people thought Snow was insane when he asked public-health officials to remove the handle of the pump. But Snow was one of the few scientists of his time who didn't believe in the miasma theory. He based his investigations on the idea that the "excretions of the sick" contained poisonous material that was passed via contaminated water from human to human, growing, multiplying, and causing illness. The theory Snow subscribed to, and what led him to demand that authorities remove the well's handle, is what we now take for granted is the true basis of infection — germ theory. But at the time, Snow was in the minority.

The premise that the worse a patient smelled, the more urgent his case made it a priority for doctors and surgeons to get to their next procedure rapidly. Things like washing your hands between patients or changing your operating gown did nothing but take up more time, so the most dedicated surgeons would go from patient to patient as quickly as possible, covered in blood and viscera. To ward off infection, they instructed the nurses to open up the windows of the surgical room to air things out.

Around the time that John Snow was removing the pump handle, another pioneering doctor was experimenting with how the idea of germ theory might change his clinical practice. Dr. Joseph Lister was a surgeon who had read the work of chemist Louis Pasteur on how wine was soured by microbes. Dr. Lister applied these concepts to his surgical practice and insisted that his surgical team employ antiseptic techniques such as hand-washing, cleaning their instruments, and cleaning the patient's skin and wounds. In the three years after Lister instituted his antiseptic practices, the death rate from infection after his surgeries went from 46 percent to 15 percent. So the next time you pick up a bottle of Listerine, know that we have Dr. Lister to thank not only for saving us from the curse of bad breath, but also for making it possible for someone to roll out of an operating room with a good chance of survival.

Despite what may seem like dramatic results, it took a very long time to get from the discovery of germ theory to the institution of universal hand-washing, the use of sterile surgical equipment, and the development of antibiotics, and it took even longer to get to our current tools of fourth-generation antibiotics and surgical equipment that's sterilized by radiation. What happened between then and now?

There are myriad small answers, of course. But they all fall into two general categories: the medical response and the public-health response. The medical response encompassed the changes in the practice of medical care, things like Lister's surgical techniques and the development of vaccines and antibiotics. The public-health response was all the ways this information changed things outside of hospitals and clinics, including the creation of practices like municipal sanitation and the pasteurization of milk.

These combined efforts were all based on a simple frame shift—that exposure to germs, not foul air, causes disease and death. Once that was accepted, people were free to get creative about limiting exposure and transmission and about ultimately treating the infections that did occur. But just as important as any individual intervention was the recognition that *both* approaches were necessary to achieve transformative change. All the antibiotics in the world won't solve the problem if people continue to dump raw sewage into the water supply. Similarly, even with the most advanced sanitation practices, some people will still get sick, so we need ways to treat infections.

I spend a lot of time with folks who ask, "What do ACEs and toxic stress have to do with me?" My medical colleagues say, "Isn't this a social problem?" And policymakers wonder, "How can we even talk about toxic stress if we don't have a cure?" The answer to all three of these questions is that understanding the mechanism of how ACEs lead to toxic stress gives us a powerful tool to shape both our medical response *and* our public-health response. And everyone has a role to play.

I believe that we are standing on the cusp of a *new* revolution, and it is every bit as consequential as the one sparked by Pasteur's discovery of germs. What's exciting is that the movement has already begun. The work that Jeannette Pai-Espinosa and Dr. Pam Cantor are doing in communities and schools is part of the ACEs public-health response. The work that Nancy Mannix and CYW are doing is part of the medical response. Right now, we are at the hand-washing stage. We have yet to develop fourth-generation antibiotics in the fight against toxic stress, but we can use the knowledge of how the stress response triggers health problems to institute some basic hygiene: Screening, trauma-informed care, and treatment. Sleep, exercise, nutrition, mindfulness, mental health, and healthy relationships — these are the equivalent of Lister dipping his instruments in carbolic acid and requiring his surgical students to wash their hands.

When we understand that the source of so many of our society's problems is exposure to childhood adversity, the solutions are as simple as reducing the dose of adversity for kids and enhancing the ability of caregivers to be buffers. From there, we keep working our way up, translating that understanding into the creation of things like more effective educational curricula and the development of blood tests that identify biomarkers for toxic stress — things that will lead to a wide range of solutions and innovations, reducing harm bit by bit, and then leap by leap. The cause of harm — whether that's microbes or childhood adversity — does not need to be totally eradicated. The revolution is in the creative application of knowledge to mitigate harm wherever it pops up. Because when you know the *mechanism*, you can use that understanding in countless ways to drastically improve the human condition. That is how you spark a revolution. You shift the frame, you change the lens, and all at once the world is revealed, *and nothing is the same.*

13

In the Rearview

IT WAS 6:00 ON a Saturday morning when my husband's cell phone rang. We were on a weekend getaway in California's wine country, so the early wake-up was both unanticipated and unwelcome. Confused and groggy, Arno rolled over and pulled the comforter over his head.

"Babe." I jostled him. "Babe, it's your phone. Who the heck is calling you?"

Arno slapped one hand on the nightstand, first found his glasses, then his phone.

"Hello?" he croaked.

An instant later he was sitting up, his voice alert and quick. "Yeah, yeah, she's here. Hang on."

He thrust the phone toward me. "It's Sarah. Evan had a stroke."

What the . . . ? As a doctor, I'm accustomed to getting calls at odd hours from relatives and friends. Occasionally it's something significant (a friend's wheezing baby) and I have substantive advice to give *(Go to the ER right away!)*. But more often it feels like I'm running an advice line for the worried well *(My two-year-old ate cat poop, what should I do?* a cousin asks. *Don't let her eat any more cat poop,* I say). So when Arno handed me the phone, the main thing going through my mind was *What the heck does she mean by stroke?* I pictured my brother falling asleep with a limb tucked under him and waking up with pins and needles or possibly coming down with a case of Bell's palsy, a scary but benign inflammation of the facial nerve that can leave half your face paralyzed for weeks to months. When I took Arno's phone from him, I was feeling more skeptical than worried.

"Sarah?"

"Hi, Nadine."

My sister-in-law's voice was eerily measured.

"I'm in the ER at UCSF. The doctors here want to do an experimental procedure. They say that it could save Evan's life, but I would have to sign a consent to be part of a clinical trial. I don't know what to do. Can you talk to the doctor and let me know what you think?"

My pulse quickened. *ER? UCSF?* What was going on?

"Sure, sure, put 'em on," I said, sliding over to perch next to Arno on the side of the bed.

Seconds later, I heard a very authoritative and slightly rushed voice on the end of the line. The tone, more than anything, sent my alarm bells ringing. I recognized it immediately. It was crisp, direct, and concise, a tone I had used many times when I stood by a patient's bed and could almost see the Grim Reaper standing on the other side. There wasn't a second to waste.

The doctor briefly introduced herself and then started explaining what the problem was and what they wanted to do. I was taking it all in, nodding and mmm-hmm-ing, until I heard the phrase "blockage of two-thirds of the distribution of the middle cerebral artery."

My whole body reeled.

"Whaaaaat?" I screamed into the phone.

I knew what that meant clinically; the thing I couldn't wrap my mind around was the fact that it was happening to my brother. It meant a huge chunk of his brain wasn't getting any blood. It meant death, most likely. Or, if we were lucky, severe disability. I pictured Evan in a wheelchair with one arm tucked into his chest like a bird with an unusable, broken wing. I pictured adult diapers and home-health aides to help turn him in bed. I pictured applesauce dribbling down the droopy side of his mouth.

I started to sob.

I could feel Arno's hand gently rubbing the small of my back. I took a deep breath and kept listening.

The doctor paused for a moment and then began again, a little more slowly at first, then picking up the pace. She laid out the survival rates for the standard treatment and explained why she thought Evan's case was a particularly good one for this new, experimental procedure. I forced myself to take it all in. She explained the risks and the potential benefits, and when she finally wrapped up and told me she was handing the phone back to my sister-in-law, I had to pull myself together. There was no way that I could let Sarah hear the distress in my voice.

"Sarah. Sounds like our best bet is to do this procedure."

I did my best to sound calm and reassuring.

"Really? Are you sure?"

"Absolutely," I answered. "It's our best shot."

Ninety minutes later, we stepped through the sliding glass door of the neurosurgical intensive care unit at UCSF. Arno carried three-year-old Kingston in his arms. We were escorted to the waiting room, where my parents and my other brothers were keeping a vigil. In the hours that we waited for the procedure to be complete, I could periodically hear the doctors and the nurses in the ICU relaying information about his case: "Forty-three-year-old male with acute stroke, non-smoker, no risk factors." The last part echoed and rattled around in my brain. *No risk factors.*

That wasn't true.

When my brothers and I were growing up, our mother suffered from paranoid schizophrenia, a severe form of mental illness that, unfortunately, went untreated for many years. As it is for most families with that legacy, the story was complicated. In our house, times of intense anxiety and stress were interwoven with moments of love and joy. My mom taught me how to hit a mean two-handed backhand in tennis and was the fiercest educational advocate anyone could imagine, always saying to me, "Get your education, girl, because once you have it no one can take that away from you!" But when it was bad . . . well, it was pretty darn bad. The problem was that we never knew which mother we were going to get. Every day after school it was a guessing game — are we coming home to happy Mom or scary Mom? Needless to say, it created an environment of repeated and unpredictable stress that marked us in different ways, both negative and positive.

That day, as I sat in the waiting room of the neurosurgical ICU, sick with worry, I couldn't help thinking about how different things might have been if Evan's ACE score had been a part of his medical history. Folks with significant ACEs are more than twice as likely to have a stroke. How could his care have been different leading up to this moment if his ACE score was treated as a biological indicator just like blood pressure or cholesterol? If we had known how ACEs are related to this particular kind of stroke, could we have modified the risk? Could this knowledge help prevent the next person like Evan from ever having a stroke? All these questions led me to the same conclusion — when it comes to ACEs, we need more research, desperately.

Fortunately for my family, the research to advance the treatment of stroke paid off. As a doctor, I don't say this lightly: the experimental procedure that saved my brother's life was nothing short of miraculous. The team at UCSF removed the clot in its entirety and restored blood flow to Evan's brain. When he woke up in the ICU, he was still extremely weak on the right side of his body, but within a few months, with intensive physical therapy, he was back to riding his bike in the Marin Headlands and playing basketball with his boys.

. . .

When we were kids, Evan adapted to the stress at home by being a total charmer. To this day, he has a natural charisma that automatically bubbles up and puts people at ease. Sometimes I still chuckle when I remember the zingy one-liners he delivered as the emcee of our wedding. He had everyone buzzing with joy and laughter. Our brother Louis wasn't so lucky. Louis and I were a year apart and looked so much alike when we were little that people often asked if we were twins. Louis was smarter than I was, and unlike me, he was actually popular in high school. But he was also sensitive. His unique combination of nature and nurture led to his own schizophrenia; he was diagnosed in 1992, when he was just seventeen years old. Two years later, he got out of my mom's car at a stoplight and walked away. We never saw him again. He's been on the national missing-persons registry ever since. Louis is what brought me to Bayview Hunters Point. I see his face, his potential, his fundamental worth in the faces of my patients.

Looking back, I can see now how I adapted to our mom's illness by becoming more attuned to those around me. For me, quickly figuring out which mom I was coming home to was the key to navigating our household. Now it's easy for me to tell when there's something going on with people by reading a whole bunch of nonverbal cues. It's kind of like a sixth sense. I would never want to repeat the distressing or unpredictable moments of my childhood, but I wouldn't wish them away either. They are a big part of what has made me who I am. Sometimes I like to think of this ability to tune in to people as my own little superpower. As a doctor, it allows me to gently ask my patients the right follow-up questions and get to the heart of the matter quickly. This has been a huge gift for me in my practice.

My adaptation to my mom's illness also delivered benefits in medical school and residency. High-adrenaline situations were where I shone. I wouldn't be surprised to hear that many of my colleagues found a place for themselves in medicine for a similar reason. Where others might have gotten overwhelmed or flustered, my brain and body were accustomed to working in heart-pounding conditions. I'll never forget the day in the pediatric ICU at Stanford when, as a second-year resident, I was charged with removing the breathing tube of a patient who had received a liver and small bowel transplant and who we believed was recovering well enough to breathe on his own. For the first few minutes, he did well and seemed stable. But after my attending physician left the room, he suddenly and unexpectedly flatlined. My mind and body went into overdrive. Every ounce of training was deployed swiftly and with precision. When my attending came rushing back in to respond to the code blue, she found me up on the bed metering out chest compressions and calling out doses of epinephrine to the nurse. When it was all over, when we got the patient's heartbeat back and he was stabilized, my attending shook her head as we took a moment to debrief on what had just happened.

"What the hell was that?" she asked.

"What do you mean? He was in asystole. The protocol says that when the patient is in asystole, you start compressions."

She laughed. "I know that. I've just never seen a resident respond so quickly and decisively before."

I shrugged. *Well, that is what the protocol says,* I thought to myself.

That otherworldly clarity, that extra level of focus and performance, is what my brothers, who are football fans, call Beast Mode. It's what the fight-or-flight response was designed for. That day, standing just outside my patient's room in the hallway of the ICU, I smiled. Secretly, I felt as powerful and agile as a running back who had just leaped over a line of defenders and into the end zone. Nadine, 1; Grim Reaper, 0. Doctors don't get to dance a shuffle like Ickey Woods of the Cincinnati Bengals when they do something they feel particularly good about, but I *might* have gone into the ladies' room and done a fist-pump in the mirror.

. . .

My experience dealing with both sides of the ACEs coin is in part what drives my work. I know that the long-term impacts of childhood adversity are not all suffering. In some people, adversity can foster perseverance, deepen empathy, strengthen the resolve to protect, and spark mini-superpowers, but in all people, it gets under our skin and into our DNA, and it becomes an important part of who we are.

I don't think people who grew up with ACEs have to "overcome" their childhoods. I don't think forgetting about adversity or blaming it is useful. The first step is taking its measure and looking clearly at the impact and risk as neither a tragedy nor a fairy tale but a meaningful reality in between. Once you understand how your body and brain are primed to react in certain situations, you can start to be proactive about how you approach things. You can identify triggers and know how to support yourself and those you love.

This is about understanding how adversity disrupts the delicate ecosystems of family and overwhelm us. It's about recognizing that when it inevitably does happen, we can use what we've learned from science to do a better job helping ourselves and one another so we can better protect our children. As parents and caregivers, we can find it hard to admit when we're struggling. It's really easy to get caught up in feeling guilty and ashamed about all the ways, both real and imagined, that we have failed our kids. But one of the things I hope you will take away from these pages is an understanding that how adversity affects you is not a referendum on your character. We don't need to play the shame game. It doesn't help.

I'm not saying that any of this is easy.

If you're someone with an ACE score of your own, learning to recognize when your stress response is getting out of whack can be hard. Taking the time and finding the resources to do self-care and get yourself on the path to healing can be even harder. If you're a parent with ACEs, or even a parent without ACEs, you have a double challenge because you have to worry about taking care of yourself *and* protecting your child. Or, as we've learned, doing the former so you can do the latter.

I learned about the powerful ability of trauma and adversity to shape who we are and how our bodies work as a physician on a quest to heal my patients, but in a sad and unexpected twist, I got to know it in a totally different way — as a mom.

I know what it's like to be an impaired parent. When I travel and speak, I often tell folks about our crazy blended family and our four beautiful boys. But that's a lie I use to make other people feel comfortable. The truth is that we have five boys. One year before Evan had his stroke, I had a medical crisis of my own. Ziggy Harris was born on January 31, 2014, at 5:51 a.m. He lived for fourteen minutes and thirty-seven seconds. The moment the nurse took him — blue and lifeless — from my arms was the single worst moment of my life.

Ziggy had been my secret friend for six beautifully anticipatory months. As any pregnant mother can understand, we were BFFs long before he took his first or last breath. He liked pineapple, *hated* the smell of cooking meat, and his favorite position was snuggling head-down on the right side of my womb. I was pretty sure he was pursuing a black belt in jujitsu based on the kicks that landed on my left rib cage. When we lost him, to say that I was a mess would be the understatement of the century.

Arno and I grieved very differently. He was focused on taking care of everyone, especially the boys. He made sure they got to school on time, that groceries were in the fridge and food was on the table. I, however, couldn't function. I couldn't take care of myself, much less anyone else.

One morning, about three days after we lost Ziggy, I got up at four thirty. I couldn't sleep. In a cruel twist of biology, my milk was coming in. All of a sudden, I couldn't stand being in the house anymore. Everything reminded me of the baby. The body pillow that I had used to support my growing belly now lay useless on the floor next to our bed. I couldn't look at it. I begged Arno to take me somewhere else. I needed to get out of the house.

Arno's face revealed a mix of deep concern and fear. It was clear that he was worried that his wife might be losing her mind.

"Babe, what are you talking about?" he asked gently. "The kids have to go to school today."

My eyes fixed on my husband. Why the fuck was he talking about the kids going to school? I needed to be away. I couldn't stand to be in that house for one. More. Minute.

"Well, if you won't take me, I'll go by my damn self!" I screamed, then I grabbed my car keys and stormed out the door, leaving my husband at home with our three sleeping children. I wanted to get out of my skin. I was hoping to drive until I found a place where it didn't hurt so much. That was a mistake. The only thing worse than being at home was being alone.

An hour later, I found myself sitting in my car in front of the Starbucks on Irving and Ninth, sobbing hysterically into the steering wheel. I had to figure out what the hell I was going to do now.

I looked up and caught my reflection in the rearview mirror. For a moment, I almost didn't recognize myself. Staring back at me in the mirror, wild-eyed, was the semblance of my mother.

Out of nowhere, there was a *tap, tap, tap* on my window.

In what I can only call an act of divine intervention, Evan was out for an early-morning run, and, of all the places in the city, he happened to be coming down Irving Street and recognized my car.

I rolled down the window.

"Are you okay?" Evan asked.

And in that moment, I realized that I wasn't. I *really* wasn't okay. I needed help.

The minute I recognized that I was unable to function, my first thought was *How do I keep this from hurting my kids?* Because of what I'd seen in my work, I knew that my falling apart didn't affect just me. I also knew that two things would be critical to getting our family through this. The first was making sure the kids had the buffering care and love they needed. The second was getting the support and care that I needed. That knowledge made all the difference in the world.

Later that day, Sarah came to stay with us. She provided the safe, stable, and nurturing environment for our children that I couldn't. She took care of the kids so that Arno could focus on taking care of me. It wasn't until that crazy morning that we figured out that he couldn't do both — we needed the village. I will never be able to express my gratitude to Evan and Sarah for being there for us and for our children during our most difficult moments.

There isn't a day that goes by that I don't think about the son that we lost. And despite my tendency toward optimism, I have struggled to find meaning in his passing. But I do recognize that we were lucky. In the moment that I was brought to my knees, I had folks I could lean on to help me get back up. That's something that I am profoundly grateful for. Sitting in my car, crying in front of Starbucks, I caught a glimpse of what it might be like to lose the ability to be the parent we all want to be. My mother didn't have the network of support that Arno and I enjoy. She also didn't have the benefit of two decades of research on toxic stress to tell her what the impacts on her children might be and what she could do to help herself and her kids. She did the best she could with what she had.

But we have more now; we know more. I believe that we can rewrite the story of adversity and break the intergenerational cycle of toxic stress. I wrote this book for all of the parents, stepparents, foster parents, grandparents, and caregivers of all stripes who are trying to figure out how to give the little people in their care the best shot in this world despite the difficulties life throws in their way and, often, despite their own histories of adversity. I wrote it for all of the children and young people in this world facing outsize challenges, and for the adults whose health is being shaped by the legacy of their childhoods. My hope is to inspire conversations — around dinner tables, in doctors' offices, at PTA meetings, in courtrooms, and at city councils. But my greatest hope is to inspire action — big and small.

Whether it's simply learning to recognize when your own stress response is activated and figuring out how to respond in a way that is healthy and not harmful to the people you love, or becoming a mentor to a child in need, or talking to your doctor, there is something that every one of us can do to change the way we, as a society, respond to ACEs.

I believe that when we each find the *courage* to look this problem in the face, we will have the power to transform not only our health, but our world.

Epilogue

IT'S 2040 AND THINGS are a little different. I'm a grandmother now (but wouldn't you know it, I still look good). I'm retired, and when I'm not putzing around in my garden, I keep busy chasing the grandkids. They are four, five, and seven, and of course I spoil them rotten, the guilty pleasure of every grandparent since the beginning of time.

Our eldest sons (the twins) are thirty-seven and I'm in love with my daughters-in-law, who both called me directly after their first prenatal appointments to tell me they had an ACE screening as part of their routine prenatal care. Even though it's standard nowadays, they know how much I still love to hear about doctors following through on the guidelines that CYW helped develop. Our boys just roll their eyes when their wives indulge me as I rattle on with my "back in the day" stories, but I know they are secretly proud each time they fill out the school forms for their kids and see the checked box that certifies each child has received an ACE screening right along with vaccinations and TB tests.

Grayboo, who now insists on being called by his proper name, teaches third grade at a public elementary school. He gives me the ACE scoop from the other side of the desk, telling me how the school incorporates ACE awareness into its teacher training. One of the first things the school makes sure of is that teachers know how to recognize symptoms of toxic stress in their students. Every morning, Gray guides his class in a Quiet Time meditation practice to help his students hit the reset button as they start the day, reinforcing the self-regulation skills they have been working on throughout the year.

Even though I'm retired, I still make time to teach at least one course on ACEs and toxic stress to first-year medical students at Stanford, where Kingston is now part of the class. We start at the beginning of the semester with the biological mechanisms, and by the end we're discussing the latest interventions for healing a disrupted neuro-endocrine-immune system.

On the public-health side of things, the movement has taken off. Two decades ago, CYW was instrumental in convening a group of advocacy and education organizations led by the American Heart Association, the American Cancer Society, and the American Lung Association and together they created a powerful public-education campaign. It started with a viral video and spun out from there — billboards, posters in doctors' offices, a Super Bowl ad, and more. Celebrities volunteer to be part of the Faces of ACEs ad campaign, and they share their stories along with the call to action: *Know your score and learn how to heal.* My sons' generation is the first to grow into adulthood without the stigma surrounding adversity. These days, having an ACE score isn't any more shameful than having a peanut allergy. But the campaign did far more than change attitudes; twenty-plus years later, we have seen a 40 percent decline in the number of Americans reporting one or more ACEs, and a 60 percent decline in the number of Americans reporting four or more ACEs. Adverse events still happen to all kinds of people, but they are no longer handed down from generation to generation to generation.

The Resilience Investment Act of 2020, which provided federal dollars for screening, treatment, and research, created a national consortium modeled after the Children's Oncology Group that is wildly successful. The double-digit decline in health-care spending allows us to reallocate dollars to national priorities in some predictable and some surprising ways. Our increased allocation to early-childhood care and education programs was a no-brainer. The big surprise came when I got a call from the U.S. State Department asking me to help advise on a new program that will work closely with other nations' governments to deploy widespread ACE screening and early intervention in high-conflict areas. This way, we can inoculate the younger generation so they will not be susceptible to induction into gangs, militias, and insurgencies. The science of toxic stress has become a powerful tool in maintaining global security. And our military also uses the latest treatments to help our troops returning from combat.

Ultimately, I help where I can, but for the most part, there isn't much for me to do. What started out as a movement has become just how people do things — basic infrastructure, standard of medical practice, common wisdom. So Arno and I spend most of our time just being grandparents. We take the grandkids to the park, we buy them things we know we shouldn't, and when I come across my grandkids chucking paper airplanes at one another, I grab my tape measure and my stopwatch and laugh when they all roll their eyes and flee before the science lesson begins — all of them, that is, except one.

Appendix 1

WHAT'S MY ACE SCORE?

Prior to your eighteenth birthday:

1. Did a parent or other adult in the household **often** . . .
    Swear at you, insult you, put you down, or humiliate you?
    or
    Act in a way that made you afraid you might be physically hurt?
    Yes   No   If yes enter 1 _____

2. Did a parent or other adult in the household **often** . . .
    Push, grab, slap, or throw something at you?
    or
    Ever hit you so hard that you had marks or were injured?
    Yes   No   If yes enter 1 _____

3. Did an adult or person at least five years older than you **ever** . . .
    Touch or fondle you or have you touch their body in a sexual way?
    or
    Attempt or actually have oral, anal, or vaginal intercourse with you?
    Yes   No   If yes enter 1 _____

4. Did you **often** feel that . . .
    No one in your family loved you or thought you were important or special?
    or
    Your family didn't look out for each other, feel close to each other, or support each other?
    Yes   No   If yes enter 1 _____

5. Did you **often** feel that . . .
    You didn't have enough to eat, had to wear dirty clothes, and had no one to protect you?
    or
    Your parents were too drunk or high to take care of you or take you to the doctor if you needed it?
    Yes   No   If yes enter 1 _____

6. Were your parents ever separated or divorced?
    Yes   No   If yes enter 1 _____

7. Was your mother or stepmother . . .
    Often pushed, grabbed, slapped, or had something thrown at her?
    or
    Sometimes or often kicked, bitten, hit with a fist, or hit with something hard?
    or
    Ever repeatedly hit over at least a few minutes or threatened with a gun or knife?
    Yes   No   If yes enter 1 _____

8. Did you live with anyone who was a problem drinker or alcoholic or who used street drugs?
    Yes   No   If yes enter 1 _____

9. Was a household member depressed or mentally ill, or did a household member attempt suicide?
    Yes   No   If yes enter 1 _____

10. Did a household member go to prison?
    Yes   No   If yes enter 1 _____

    Now add up your "Yes" answers: _____

    This is your ACE Score.

Appendix 2

CYW ADVERSE CHILDHOOD EXPERIENCES QUESTIONNAIRE
(ACE-Q) CHILD

To Be Completed by Parent/Caregiver

Today's Date: _____

Child's Name: _____   Date of Birth: _____

Your Name: _____   Relationship to Child: _____

Many children experience stressful life events that can affect their health and well-being. The results from this questionnaire will assist your child's doctor in assessing his or her health and determining guidance. Please read the statements below. Count the number of statements that apply to your child and write the total number in the box provided.

Please DO NOT mark or indicate which specific statements apply to your child.

1) Of the statements in Section 1, HOW MANY apply to your child? Write the total number in the box. ☐

Section 1. At any point since your child was born . . .

- Your child's parents or guardians were separated or divorced.
- Your child lived with a household member who served time in jail or prison.
- Your child lived with a household member who was depressed, mentally ill, or attempted suicide.
- Your child saw or heard household members hurt or threaten to hurt each other.
- A household member swore at, insulted, humiliated, or put down your child in a way that scared your child, OR a household member acted in a way that made your child afraid that she or he might be physically hurt.
- Someone touched your child's private parts or asked your child to touch their private parts in a sexual way.
- More than once, your child went without food, clothing, or a place to live, or had no one to protect her or him.
- Someone pushed, grabbed, slapped, or threw something at your child, OR your child was hit so hard that your child was injured or had marks.
- Your child lived with someone who had a problem with drinking or using drugs.
- Your child often felt unsupported, unloved, or unprotected.

2) Of the statements in Section 2, HOW MANY apply to your child? Write the total number in the box. ☐

Section 2. At any point since your child was born . . .

- Your child was in foster care.
- Your child experienced harassment or bullying at school.
- Your child lived with a parent or guardian who died.
- Your child was separated from her or his primary caregiver through deportation or immigration.
- Your child had a serious medical procedure or life-threatening illness.
- Your child often saw or heard violence in the neighborhood or in her or his school neighborhood.
- Your child was often treated badly because of race, sexual orientation, place of birth, disability, or religion.

Acknowledgments

I must begin by thanking my patients and the families who have shared their lives with me and entrusted me with the care of their greatest treasures — their children. I am also deeply grateful to the community of Bayview Hunters Point for embracing me, supporting me, and going on this learning journey with me. Special thanks to Dwayne Jones for his guidance, for vouching for me, and for throwing the weight of the mayor's office behind the Bayview clinic.

Though it has long been a dream of mine, I never imagined that I would actually write a book. There is an old saying that "you have to see it to be it." I'm grateful to my dear friends Kathleen Kelly Janus and Anja Manuel for fearlessly putting their voices into the world and then encouraging me to do the same. I also have to thank Faye Morrison, my fifth- and sixth-grade teacher at Ohlone Elementary School in Palo Alto, for nurturing my love of reading and writing.

Thank you to Rachel and Zara for lovingly caring for my children so that I can care for the children of others.

Paul and Daisy Soros supported my medical education and gave me the freedom to practice where my heart (as opposed to my student loans) led me. Stan Heginbotham and Warren Ilchman at the Paul and Daisy Soros Fellowships for New Americans encouraged me to "compose a life" and to get out into the world and learn by doing.

Credit goes to the National Institutes of Health for supporting my public-health education and research training.

Thanks to Martin Brotman, Steve Lockheart, and Terry Giovannini at CPMC who believed in my crazy dream of opening a clinic in Bayview Hunters Point when I was fresh out of residency.

I've also benefited from the wise mentorship of Cheryl Polk, Ann O'Leary, Jennifer Siebel Newsom, Esta Soler, Suzy Loftus, Lenore Anderson, Jennifer Pitts, George Halvorson, Geoff Canada, Bryan Stevenson, and Kamala Harris.

The first step to healing toxic stress is understanding that it exists in the first place. Thank you to Jamie Redford, Ashley Judd, and Anna Deavere Smith for shouting it from the rooftops.

I first met Paul Tough at a conference in New York in 2009. When I heard that he worked (at the time) for the *New York Times Magazine*, I launched into a forty-five-minute monologue on ACEs and toxic stress. I'm so thankful that he didn't run for the hills but rather listened and amplified.

All of the research and science in the book is the product of the tireless efforts of researchers and physicians who have come before me and who continue to make important advances in our understanding of toxic stress and its treatment. There are too many to name, but I want to share how much they inspire me and how grateful I am to those who have laid the scientific foundation for this field. I owe particular thanks to Monica Singer, Sarah Hemmer, Whitney Clarke, Todd Renschler, Lisa Gutierrez Wang, Susan Briner, Denise Dowd, Andy Garner, Eva Ihle, Sheila Ohlsson Walker, Pamela Cantor, Jack Shonkoff, Tom Boyce, Nancy Adler, Roy Wade, Mark Raines, Alicia Lieberman, Rob Anda, Vince Felitti, and Victor Carrion, all of whom have greatly influenced my thinking and shaped my approach to identifying and treating ACEs and toxic stress.

I owe a great deal to Justin Sherman, my leadership coach, whose patient guidance kept me going when I was ready to give up.

All of the data in this book was meticulously assembled by Debby Oh with help from Sukhdip Purewal and other members of our superb research team at the Center for Youth Wellness, including Monica Bucci and Kadiatou Koita. These women have an unmatched commitment to rigor and precision. In addition, I want to thank our exceptional teams at the Center for Youth Wellness and the Bayview Child Health Center as well as our stellar board of directors (past and present), our leadership council, and our community advisory council. It is my great joy to work side by side with these thoughtful and dedicated people who manage to demonstrate the healing power of individual relationships on a day-to-day basis and share that vision to improve the health and lives of millions.

I feel very fortunate to be supported by the brilliant team at Houghton Mifflin Harcourt. I deeply appreciate the efforts of Tracy Roe, my sharp and hilariously funny copyeditor, and Deanne Urmy, whose engaged questions and skillful edits have made this book better than I could have hoped.

Credit for encouraging me to write this book goes to Doug Abrams, my literary agent, whose bold spirit ignites boldness in others. I owe much to him and to the wonderful team at Idea Architects, including Lara Love Hardin and, most especially, my amazing collaborator Lauren Hamlin. Thank you, Lauren, for your creativity, your diligence, your partnership, and your wicked sense of humor.

Everything that I have accomplished in my life has been because someone believed and invested in me. I would not be where I am if it weren't for the generosity of a few folks who made early bets on me and my team and who nurtured me along the way, including George Sarlo, Elaine Gold, Tom and JaMel Perkins, John and Lisa Pritzker, Bob Ross and the folks at the California Endowment, Russ and Beth Siegelman, Warren Browner at CPMC, Barbara Picower, Jaquelline Fuller and the team at Google.org, Daniel Lurie and the team at Tipping Point, and Ruth Shaber and the team at the Tara Health Foundation. Special thanks to Dr. Shaber for reading drafts and offering unflinching insights and suggestions to make this book the best it could be.

In addition, I want to thank my patients, colleagues, friends, and family who generously shared their stories for this book. My great hope is that their histories provide the soil from which the seeds of healing can grow.

My deepest gratitude goes to my family: my mom, dad, brothers, sister-in-law, cousins, aunties, uncles, and the whole extended clan of Jamaicans both in the United States and back a yaad. They have been my village, and they exemplify the word *resilience*.

Our four boys — Petros, Paulos, Kingston, and Gray — provide me with the joy and inspiration to give my best every day for the next generation.

Finally, there are no words to adequately honor my phenomenal husband. The greatest luck of my life was meeting Arno Harris. He is a wellspring of love, connection, kindness, joy, patience, and hilarity in my life. Plus he's whip-smart and pretty darned hot. I owe him a deep debt of gratitude, not only for reading countless drafts and offering invaluable suggestions and edits, but also for taking on more of the diaper changes, pickups, drop-offs, meal preparations, bathtimes, and bedtime book reading than will ever be fair so that I could stay up late and get up early to make this book come into the world.

Notes

1. Something's Just Not Right

page

5 *The diagnostic criteria:* "Attention-Deficit / Hyperactivity Disorder (ADHD),"
Centers for Disease Control and Prevention, October 5, 2016, https://www.cdc.gov/
ncbddd/adhd/diagnosis.html.

6 *a rare disorder:* Mark Deneau et al., "Primary Sclerosing Cholangitis, Autoimmune Hepatitis, and Overlap in Utah Children: Epidemiology and Natural History,"
Hepatology 58, no. 4 (2013): 1392–1400.

8 2004 Community Health Assessment: *2004 Community Health Assessment:
Building a Healthier San Francisco* (December 2004).
leading cause of early death: Ibid., 117.
Right next to Bayview: Ibid., 42.

9 *In a documentary: Take This Hammer,* directed by Richard O. Moore, National
Education Television, 1963, https://diva.sfsu.edu/bundles/187041.

11 *The Broad Street area:* Judith Summers, *Soho: A History of London's Most
Colourful Neighborhood* (London: Bloomsbury, 1989), 113–17.
By canvassing the residents of the Broad Street neighborhood: Steven John-
son, *The Ghost Map: The Story of London's Most Terrifying Epidemic — and How It
Changed Science, Cities, and the Modern World* (New York: Riverhead Books, 2006),
195–96.

2. To Go Forward, Go Back

19 *What he found was:* T. B. Hayes and T. H. Wu, "The Role of Corticosterone in
Anuran Metamorphosis and Its Potential Role in Stress-Induced Metamorphosis,"
Netherlands Journal of Zoology 45 (1995): 107–9.

20 *other unexpected negative effects:* Ibid.

24 *ten million adults have this condition:* James Norman, "Hypothyroidism (Underactive Thyroid Part 1: Too Little Thyroid Hormone)," Vertical Health LLC, http://

www.endocrineweb.com/conditions/thyroid/hypothyroidism-too-little-thyroid-hormone.

3. Forty Pounds

29 *a clinical protocol designed:* Child Sexual Abuse Task Force and Research and Practice Core, National Child Traumatic Stress Network, *How to Implement Trauma-Focused Cognitive Behavioral Therapy* (Durham, N.C.:: National Center for Child Traumatic Stress, 2004).

30 *It was a 1998 article:* Vincent J. Felitti et al., "Relationship of Childhood Abuse and Household Dysfunction to Many of the Leading Causes of Death in Adults: The Adverse Childhood Experiences (ACE) Study," *American Journal of Preventive Medicine* 14, no. 4 (1998): 245–58.
    Felitti and Anda's research: Vincent J. Felitti, "Belastungen in der Kindheit und Gesundheit im Erwachsenenalter: die Verwandlung von Gold in Blei," *Zeitschrift für psychosomatische Medizin und Psychotherapie* 48 (2002): 359–69.

38 *ACEs were astonishingly common:* Ibid.
    four or more *categories of ACEs:* One example of statistical occurrence of disease associated with four or more ACEs is given here; some researchers use three or more ACEs as a benchmark when compiling statistical risk of associated disease.

ACE STUDY FINDINGS

In comparison to those reporting no ACEs, individuals with 4+ ACEs had significantly greater odds of reporting . . .

Ischemic heart disease	2.2
Any cancer	1.9
Chronic bronchitis or emphysema (COPD)	3.9
Stroke	2.4
Diabetes	1.6
Ever attempting suicide	12.2
Severe obesity	1.6
Two or more weeks of depressed mood in the past year	4.6
Ever using illicit drugs	4.7
Ever injecting drugs	10.3
Current smoking	2.2
Ever having a sexually transmitted disease	2.5

SOURCE: Felitti, 1998

41 *50 percent of increased likelihood:* Maxia Dong et al., "Insights into Causal Pathways for Ischemic Heart Disease," *Circulation* 110, no. 13 (2004): 1761–66; Maxia Dong et al., "Adverse Childhood Experiences and Self-Reported Liver Disease: New Insights into the Causal Pathway," *Archives of Internal Medicine* 163, no. 16 (2003): 1949–56.

4. The Drive-By and the Bear

49 *Here are the main players:* Every human brain has two hippocampi and two amygdalae. Although they are dual structures, for simplicity's sake, I refer to them as singular.

50 *A graph of the response:* Cecilio Álamo, Francisco López-Muñoz, and Javier Sánchez-García, "Mechanism of Action of Guanfacine: A Postsynaptic Differential Approach to the Treatment of Attention Deficit Hyperactivity Disorder (ADHD)," *Actas Esp Psiquiatr* 44, no. 3 (2016): 107–12.

51 *a sort of stress thermostat:* Monica Bucci et al., "Toxic Stress in Children and Adolescents," *Advances in Pediatrics* 63, no. 1 (2016): 403–28.

52 *In a 2009 study:* Jacqueline Bruce et al., "Morning Cortisol Levels in Preschool–Aged Foster Children: Differential Effects of Maltreatment Type," *Developmental Psychobiology* 51, no. 1 (2009): 14–23.
   foster kids showed dysregulated cortisol levels: Ibid., 19.

54 *the council described:* National Scientific Council on the Developing Child (2005/2014), "Excessive Stress Disrupts the Architecture of the Developing Brain: Working Paper No. 3," updated edition, https://developingchild.harvard.edu/resources/wp3.

5. Dynamic Disruption

58 *Carrion and his team:* Victor G. Carrion et al., "Decreased Prefrontal Cortical Volume Associated with Increased Bedtime Cortisol in Traumatized Youth," *Biological Psychiatry* 68, no. 5 (2010): 491–93.

61 *life expectancy of individuals with ACE scores:* David W. Brown et al., "Adverse Childhood Experiences and the Risk of Premature Mortality," *American Journal of Preventive Medicine* 37, no. 5 (2009): 389–96.

63 *hyperthyroidism among refugees:* Salam Ranabir and K. Reetu, "Stress and Hormones," *Indian Journal of Endocrinology and Metabolism* 15, no. 1 (2011): 18–22.

64 *since 1825:* Ibid.

65 *Guanfacine targets specific circuits:* Cecilio Álamo, Francisco López-Muñoz, and Javier Sánchez-García, "Mechanism of Action of Guanfacine: A Postsynaptic Differential Approach to the Treatment of Attention Deficit Hyperactivity Disorder (ADHD)," *Actas Españolas de Psiquiatría* 44, no. 3 (2016): 107–12.

66 *a child's brain:* "Five Numbers to Remember About Early Childhood Development," last updated April 2017, https://developingchild.harvard.edu/resources/five-numbers-to-remember-about-early-childhood-development/#note.

67 *Romanian orphanages:* Nim Tottenham et al., "Prolonged Institutional Rearing Is Associated with Atypically Large Amygdala Volume and Difficulties in Emotion Regulation," *Developmental Science* 13, no. 1 (2010): 46–61.

70 *newly incarcerated women:* Ranabir and Reetu, "Stress and Hormones," 18.

73 *Jerker Karlén and his colleagues:* Jerker Karlén et al., "Early Psychosocial Exposures, Hair Cortisol Levels, and Disease Risk," *Pediatrics* 135, no. 6 (2015): e1450–e1457.
   Research findings show: Shanta R. Dube et al., "Cumulative Childhood Stress

and Autoimmune Diseases in Adults," *Psychosomatic Medicine* 71, no. 2 (2009): 243–50.

74   *Researchers in Dunedin, New Zealand:* Andrea Danese et al., "Childhood Maltreatment Predicts Adult Inflammation in a Life-Course Study," *Proceedings of the National Academy of Sciences* 104, no. 4 (2007): 1319–24.
four different markers of inflammation: Ibid., 1320.

6. Lick Your Pups!

77   *Charlene brought her daughter:* Todd S. Renschler et al., "Trauma-Focused Child-Parent Psychotherapy in a Community Pediatric Clinic: A Cross-Disciplinary Collaboration," in *Attachment-Based Clinical Work with Children and Adolescents*, ed. J. Bettmann and D. Demetri Friedman (New York: Springer, 2013), 115–39.

78   *over one million new neural connections:* Center on the Developing Child, "Five Numbers to Remember About Early Childhood Development (Brief)," updated April 2017, www.developingchild.harvard.edu.

81   *two groups of rat mothers:* Dong Liu et al., "Maternal Care, Hippocampal Glucocorticoid Receptors, and Hypothalamic-Pituitary-Adrenal Responses to Stress," *Science* 277, no. 5332 (1997): 1659–62.
high-licker-leads-to-low-stress effect: Michael J. Meaney, "Maternal Care, Gene Expression, and the Transmission of Individual Differences in Stress Reactivity Across Generations," *Annual Review of Neuroscience* 24, no. 1 (2001): 1161–92.

84   *lifelong changes in the stress response:* Ian Weaver et al., "Epigenetic Programming by Maternal Behavior," *Nature Neuroscience* 7, no. 8 (2004): 847–54.

88   *childhood adversity predicts shorter telomeres:* Gene H. Brody et al., "Prevention Effects Ameliorate the Prospective Association Between Nonsupportive Parenting and Diminished Telomere Length," *Prevention Science* 16, no. 2 (2015): 171–80.
U.S. Health and Retirement Study: Eli Puterman et al., "Lifespan Adversity and Later Adulthood Telomere Length in the Nationally Representative US Health and Retirement Study," *Proceedings of the National Academy of Sciences* 113, no. 42 (2016): e6335–e6342.

89   *those with PTSD had shorter telomeres:* Aoife O'Donovan et al., "Childhood Trauma Associated with Short Leukocyte Telomere Length in Posttraumatic Stress Disorder," *Biological Psychiatry* 70, no. 5 (2011): 465–71.

91   *between 55 and 62 percent:* Leah K. Gilbert et al., "Childhood Adversity and Adult Chronic Disease: An Update from Ten States and the District of Columbia, 2010," *American Journal of Preventive Medicine* 48, no. 3 (2015): 345–49.
The states with the highest rates: Christina D. Bethell et al., "Adverse Childhood Experiences: Assessing the Impact on Health and School Engagement and the Mitigating Role of Resilience," *Health Affairs* 33, no. 12 (2014): 2106–15.

7. The ACE Antidote

101   *five separate randomized trials:* Alicia F. Lieberman, Patricia Van Horn, and Chandra Ghosh Ippen, "Toward Evidence-Based Treatment: Child-Parent Psycho-

therapy with Preschoolers Exposed to Marital Violence," *Journal of the American Academy of Child and Adolescent Psychiatry* 44, no. 12 (2005): 1241–48; Alicia F. Lieberman, Chandra Ghosh Ippen, and Patricia Van Horn, "Child-Parent Psychotherapy: 6-Month Follow-Up of a Randomized Controlled Trial," *Journal of the American Academy of Child and Adolescent Psychiatry* 45, no. 8 (2006): 913–18; Alicia F. Lieberman, Donna R. Weston, and Jeree H. Pawl, "Preventive Intervention and Outcome with Anxiously Attached Dyads," *Child Development* 62, no. 1 (1991): 199–209; Sheree L. Toth et al., "The Relative Efficacy of Two Interventions in Altering Maltreated Preschool Children's Representational Models: Implications for Attachment Theory," *Development and Psychopathology* 14, no. 4 (2002): 877–908; Dante Cicchetti, Fred A. Rogosch, and Sheree L. Toth, "Fostering Secure Attachment in Infants in Maltreating Families Through Preventive Interventions," *Development and Psychopathology* 18, no. 3 (2006): 623–49.

102 *infants of depressed moms:* Roseanne Armitage et al., "Early Developmental Changes in Sleep in Infants: The Impact of Maternal Depression," *Sleep* 32, no. 5 (2009): 693–96.

for just about every sleep disorder: Sandhya Kajeepeta et al., "Adverse Childhood Experiences Are Associated with Adult Sleep Disorders: A Systematic Review," *Sleep Medicine* 16, no. 3 (2015): 320–30; Karolina Koskenvuo et al., "Childhood Adversities and Quality of Sleep in Adulthood: A Population-Based Study of 26,000 Finns," *Sleep Medicine* 11, no. 1 (2010): 17–22; Yan Wang et al., "Childhood Adversity and Insomnia in Adolescence," *Sleep Medicine* 21 (2016): 12–18.

Nighttime sleep plays: Michael R. Irwin, "Why Sleep Is Important for Health: A Psychoneuroimmunology Perspective," *Annual Review of Psychology* 66 (2015): 143–72

increased levels of stress hormones: Ibid.

The downstream effect: Ibid.

103 *reduced effectiveness:* Ibid

growth and development: Ibid.

105 *high doses of maternal stress:* Megan V. Smith, Nathan Gotman, and Kimberly A. Yonkers, "Early Childhood Adversity and Pregnancy Outcomes," *Maternal and Child Health Journal* 20, no. 4 (2016): 790–98; Inge Christiaens, Kathleen Hegadoren, and David M. Olson, "Adverse Childhood Experiences Are Associated with Spontaneous Preterm Birth: A Case-Control Study," *BMC Medicine* 13, no. 1 (2015): 124; Vanessa J. Hux, Janet M. Catov, and James M. Roberts, "Allostatic Load in Women with a History of Low Birth Weight Infants: The National Health and Nutrition Examination Survey," *Journal of Women's Health* 23, no. 12 (2014): 1039–45; Alice Han and Donna E. Stewart, "Maternal and Fetal Outcomes of Intimate Partner Violence Associated with Pregnancy in the Latin American and Caribbean Region," *International Journal of Gynecology and Obstetrics* 124, no. 1 (2014): 6–11.

109 *acts like Miracle-Gro:* Aaron Kandola et al., "Aerobic Exercise as a Tool to Improve Hippocampal Plasticity and Function in Humans: Practical Implications for Mental Health Treatment," *Frontiers in Human Neuroscience* 10 (2016): 179–88; Nuria Garatachea et al., "Exercise Attenuates the Major Hallmarks of Aging," *Rejuvenation Research* 18, no. 1 (2015): 57–89.

110 *reduce the presence of inflammatory cytokines:* Eduardo Ortega, "The 'Bioregu-

latory Effect of Exercise' on the Innate/Inflammatory Responses," *Journal of Physiology and Biochemistry* 72, no. 2 (2016): 361–69.

Eating foods that are high in: Cristiano Correia Bacarin et al., "Postischemic Fish Oil Treatment Restores Long-Term Retrograde Memory and Dendritic Density: An Analysis of the Time Window of Efficacy," *Behavioural Brain Research* 311 (2016): 425–39; A. L. Dinel et al., "Dairy Fat Blend Improves Brain DHA and Neuroplasticity and Regulates Corticosterone in Mice," *Prostaglandins, Leukotrienes and Essential Fatty Acids* (PLEFA) 109 (2016): 29–38; Javier Romeo et al., "Neuroimmunomodulation by Nutrition in Stress Situations," *Neuroimmunomodulation* 15, no. 3 (2008): 165–69; Lianne Hoeijmakers, Paul J. Lucassen, and Aniko Korosi, "The Interplay of Early-Life Stress, Nutrition, and Immune Activation Programs Adult Hippocampal Structure and Function," *Frontiers in Molecular Neuroscience* 7 (2014); Kit-Yi Yam et al., "Early-Life Adversity Programs Emotional Functions and the Neuroendocrine Stress System: The Contribution of Nutrition, Metabolic Hormones and Epigenetic Mechanisms," *Stress* 18, no. 3 (2015): 328–42; Aisha K. Yousafzai, Muneera A. Rasheed, and Zulfiqar A. Bhutta, "Annual Research Review: Improved Nutrition–A Pathway to Resilience," *Journal of Child Psychology and Psychiatry* 54, no. 4 (2013): 367–77.

a diet high in refined sugar: Janice K. Kiecolt-Glaser, "Stress, Food, and Inflammation: Psychoneuroimmunology and Nutrition at the Cutting Edge," *Psychosomatic Medicine* 72, no. 4 (2010): 365.

Elizabeth Blackburn and Elissa Epel's research showed: Elizabeth Blackburn and Elissa Epel, *The Telomere Effect: A Revolutionary Approach to Living Younger, Healthier, Longer* (New York: Grand Central Publishing, 2017).

111 *Dr. John Zamarra:* John W. Zamarra et al., "Usefulness of the Transcendental Meditation Program in the Treatment of Patients with Coronary Artery Disease," *American Journal of Cardiology* 77, no. 10 (1996): 867–70.

arterial-wall thickness: Amparo Castillo-Richmond et al., "Effects of Stress Reduction on Carotid Atherosclerosis in Hypertensive African Americans," *Stroke* 31, no. 3 (2000): 568–73.

112 *In another study:* L. E. Carlson et al., "Mindfulness-Based Stress Reduction in Relation to Quality of Life, Mood, Symptoms of Stress and Levels of Cortisol, Dehydroepiandrosterone Sulfate (DHEAS) and Melatonin in Breast and Prostate Cancer Outpatients," *Psychoneuroendocrinology* 29, no. 4 (2004): 448–74, doi: 10.1016/s0306-4530(03)00054-4.

113 *more than sixty thousand young people:* Michael T. Baglivio et al., "The Prevalence of Adverse Childhood Experiences (ACE) in the Lives of Juvenile Offenders," *Journal of Juvenile Justice* 3, no. 2 (2014): 1.

9. Sexiest Man Alive

136 *a brain wave test:* Jean Koch, *Robert Guthrie—the PKU Story: A Crusade Against Mental Retardation* (Pasadena, CA: Hope Publishing, 1997), 155–56.

IQ as twenty-five: Ibid.

137 *Though the test was accurate:* Jason Gonzalez and Monte S. Willis, "Robert Guthrie, MD, PhD," *Laboratory Medicine* 40, no. 12 (2009): 748–49, http://labmed. oxfordjournals.org/content/40/12/748.
four hundred thousand infants from twenty-nine states: Ibid.

143 *In babies, exposure to ACEs:* Anna E. Johnson et al., "Growth Delay as an Index of Allostatic Load in Young Children: Predictions to Disinhibited Social Approach and Diurnal Cortisol Activity," *Development and Psychopathology* 23, no. 3 (2011): 859–71; Marcus Richards and M. E. J. Wadsworth, "Long-Term Effects of Early Adversity on Cognitive Function," *Archives of Disease in Childhood* 89, no. 10 (2004): 922–27; Meghan L. McPhie, Jonathan A. Weiss, and Christine Wekerle, "Psychological Distress as a Mediator of the Relationship Between Childhood Maltreatment and Sleep Quality in Adolescence: Results from the Maltreatment and Adolescent Pathways (MAP) Longitudinal Study," *Child Abuse & Neglect* 38, no. 12 (2014): 2044–52.
School-age children: Paul Lanier et al., "Child Maltreatment and Pediatric Health Outcomes: A Longitudinal Study of Low-Income Children," *Journal of Pediatric Psychology* 35, no. 5 (2009): 511–22; Anita L. Kozyrskyj et al., "Continued Exposure to Maternal Distress in Early Life Is Associated with an Increased Risk of Childhood Asthma," *American Journal of Respiratory and Critical Care Medicine* 177, no. 2 (2008): 142–47; Peter A. Wyman et al., "Association of Family Stress with Natural Killer Cell Activity and the Frequency of Illnesses in Children," *Archives of Pediatrics & Adolescent Medicine* 161, no. 3 (2007): 228–34; Miriam J. Maclean, Catherine L. Taylor, and Melissa O'Donnell, "Pre-Existing Adversity, Level of Child Protection Involvement, and School Attendance Predict Educational Outcomes in a Longitudinal Study," *Child Abuse & Neglect* 51 (2016): 120–31; Timothy T. Morris, Kate Northstone, and Laura D. Howe, "Examining the Association Between Early Life Social Adversity and BMI Changes in Childhood: A Life Course Trajectory Analysis," *Pediatric Obesity* 11, no. 4 (2016): 306–12; Gregory E. Miller and Edith Chen, "Life Stress and Diminished Expression of Genes Encoding Glucocorticoid Receptor and B2-Adrenergic Receptor in Children with Asthma," *Proceedings of the National Academy of Sciences* 103, no. 14 (2006): 5496–5501; Nadine J. Burke et al., "The Impact of Adverse Childhood Experiences on an Urban Pediatric Population," *Child Abuse and Neglect* 35, no. 6 (2011): 408–13.

144 *irreversible changes:* Zulfiqar A. Bhutta, Richard L. Guerrant, and Charles A. Nelson, "Neurodevelopment, Nutrition, and Inflammation: The Evolving Global Child Health Landscape," *Pediatrics* 139, supplement 1 (2017): S12–S22.

146 *windows of neuroplasticity:* Cheryl L. Sisk and Julia L. Zehr, "Pubertal Hormones Organize the Adolescent Brain and Behavior," *Frontiers in Neuroendocrinology* 26, no. 3 (2005): 163–74; Pilyoung Kim, "Human Maternal Brain Plasticity: Adaptation to Parenting," *New Directions for Child and Adolescent Development* 2016, no. 153 (2016): 47–58.
All of these hormones: Ibid.

149 *stressors at the community level:* Roy Wade et al., "Household and Community-Level Adverse Childhood Experiences and Adult Health Outcomes in a Diverse Urban Population," *Child Abuse and Neglect* 52 (2016): 135–45.

151 *haunt her for the rest of her life:* AHRQ Patient Safety, *TeamSTEPPS: Sue Sheridan on Patient and Family Engagement,* YouTube video, posted April 2015, https://www.youtube.com/watch?v=Hgug-ShbqDs.

152 *Thanks in part to Sue:* Susan Carr, "Kernicterus: A Diagnosis Lost and Found," *Newsletter of the Society to Improve Diagnosis in Medicine* 2, no. 2 (2015): 1–3.

10. Maximum-Strength Bufferin'

168 *The great news is:* Academy of Integrative Health and Medicine, "What Is Integrative Medicine?," https://www.aihm.org/about/what-is-integrative-medicine/.

169 *And it buffers the stress response:* I. D. Neumann et al., "Brain Oxytocin Inhibits Basal and Stress-Induced Activity of the Hypothalamo-Pituitary-Adrenal Axis in Male and Female Rats: Partial Action Within the Paraventricular Nucleus," *Journal of Neuroendocrinology* 12, no. 3 (2000): 235–44; Camelia E. Hostinar and Megan R. Gunnar, "Social Support Can Buffer Against Stress and Shape Brain Activity," *AJOB Neuroscience* 6, no. 3 (2015): 34–42.

11. The Rising Tide

181 *demonstrating that Alberta:* Keith S. Dobson and Dennis Pusch, "The ACEs Alberta Program: Phase Two Results — A Primary Care Study of ACEs and Their Impact on Adult Health," presentation, November 2015.
the list goes on: Ibid.

12. Listerine

206 *COG membership includes:* Maura O'Leary et al., "Progress in Childhood Cancer: 50 Years of Research Collaboration, a Report from the Children's Oncology Group," *Seminars in Oncology* 35, no. 5 (2008): 484–93.

207 *In 1955, the National Cancer Institute (NCI):* "SWOG: History," SWOG, http://swog.org/visitors/history.asp.
Congress allocated five million dollars: Ronald Piana, "The Evolution of U.S. Cooperative Group Trials: Publicly Funded Cancer Research at a Crossroads," *ASCO Post,* March 15, 2014, http://www.ascopost.com/issues/march-15-2014/the-evolution-of-us-cooperative-group-trials-publicly-funded-cancer-research-at-a-crossroads/.

209 *46 percent to 15 percent:* C. N. Trueman, "Joseph Lister," History Learning Site, www.historylearningsite.co.uk.

Epilogue

223 *Quiet Time meditation practice:* David Lynch Foundation, "The Quiet Time Program: Restoring a Positive Culture of Academics and Well-Being in High-Need School Communities," https://www.davidlynchfoundation.org/pdf/Quiet-Time-Brochure.pdf.

Index